Metallic Functional Materials: Development and Applications

Metallic Functional Materials: Development and Applications

Guest Editors

Changlong Tan
Kun Zhang
Yan Feng

MDPI

Basel • Beijing • Wuhan • Barcelona • Belgrade • Novi Sad • Cluj • Manchester

Guest Editors
Changlong Tan
School of Materials Science
and Chemical Engineering
Harbin University of Science
and Technology
Harbin
China

Kun Zhang
Institute of Metal Research
Chinese Academy of Sciences
Shenyang
China

Yan Feng
State Key Laboratory of
Solidification Processing
Northwestern
Polytechnical University
Xi'an
China

Editorial Office
MDPI AG
Grosspeteranlage 5
4052 Basel, Switzerland

This is a reprint of the Special Issue, published open access by the journal *Metals* (ISSN 2075-4701), freely accessible at: https://www.mdpi.com/journal/metals/special_issues/H6WSB2747E.

For citation purposes, cite each article independently as indicated on the article page online and as indicated below:

Lastname, A.A.; Lastname, B.B. Article Title. *Journal Name* **Year**, *Volume Number*, Page Range.

ISBN 978-3-7258-6946-6 (Hbk)
ISBN 978-3-7258-6947-3 (PDF)
https://doi.org/10.3390/books978-3-7258-6947-3

Contents

About the Editors

Changlong Tan

Changlong Tan is a Professor at the School of Materials Science and Chemical Engineering, Harbin University of Science and Technology. His research interests focus on metallic functional materials, specifically shape memory alloys, martensitic transformations, and the application of first-principles calculations in material design. He has successfully utilized computation-guided strategies and machine learning to develop ferromagnetic shape memory alloys with giant magnetocaloric effects and excellent mechanical properties. Prof. Tan actively contributes to the field through the exploration of high-performance elastocaloric materials and serves as a Guest Editor for multiple Special Issues on functional metals.

Kun Zhang

Kun Zhang is a researcher at the Institute of Metal Research, Chinese Academy of Sciences. He specializes in shape memory alloys, caloric materials, and phase transition kinetics. His work integrates machine learning with physical descriptors to optimize functional properties in metastable metallic systems, specifically focusing on the cyclic stability and elastocaloric cooling performance of Heusler alloys. His research has provided deep insights into the microstructural evolution and thermodynamic properties of novel alloy systems, contributing to the advancement of solid-state refrigeration technologies.

Yan Feng

Yan Feng is an Associate Professor at the State Key Laboratory of Solidification Processing, Northwestern Polytechnical University. Her research encompasses shape memory alloys, microstructure analysis, and additive manufacturing. She has investigated the effects of laser solid forming on the phase transformation of NiTi alloys and the enhancement of elastocaloric effects in magnetic shape memory alloys through element doping. She has previously served as a Special Assistant Professor at Osaka University and has been recognized with provincial awards for her technological inventions in material forming.

metals MDPI

Editorial

Metallic Functional Materials: Development and Applications

Changlong Tan [1,*], Kun Zhang [2] and Yan Feng [3,*]

[1] School of Materials Science and Chemical Engineering, Harbin University of Science and Technology, Harbin 150000, China
[2] Institute of Metal Research, Chinese Academy of Sciences, Shenyang 110016, China; kzhang@imr.ac.cn
[3] State Key Laboratory of Solidification Processing, Northwestern Polytechnical University, Xi'an 710072, China
* Correspondence: changlongtan@hrbust.edu.cn (C.T.); yanfeng@nwpu.edu.cn (Y.F.)

1. Introduction and Scope

Metallic functional materials have become a strategic focus in contemporary material research, driven by the rising demand for intelligent, adaptive, and high-performance systems across energy conversion, aerospace actuation, biomedical devices, and next-generation flexible or micro-electromechanical technologies. Beyond their structural roles, these materials—ranging from shape memory alloys to magnetic, thermoelectric, high-entropy, and metastable metallic systems—integrate tunable physical responses with mechanical reliability, bridging fundamental metallurgy and device-level innovation.

This Special Issue of *Metals*, brings together contributions that exemplify the rapid expansion of this field. The selected papers highlight how integrated computational approaches, materials-genome strategies, and machine-learning-assisted design accelerate the exploration of complex compositional spaces; how advanced processing routes such as additive manufacturing, rapid solidification, and severe plastic deformation stabilize novel metastable phases; and how operando characterization techniques deepen our understanding of phase transitions, defect kinetics, and multi-field coupling phenomena. Equally important are studies designed to propel functional metals toward real-world applications, including high-temperature actuators, precision damping elements, biomedical micro-components, and energy-harvesting or thermal management devices designed for stringent operational environments.

Several scientific trends clearly emerge across this Special Issue. Microstructure has become an active design variable: hierarchical interfaces, controlled disorder, gradient architectures, and nanoscale phase coexistence are now systematically engineered to achieve targeted functionalities. Predictive theory and data-driven modeling are reshaping alloy discovery, replacing traditional trial-and-error strategies with mechanism-guided, high-throughput exploration. Meanwhile, application-driven constraints are forcing a deeper examination of reliability, cyclic stability, fatigue resistance, and environmental interactions—issues critical in long-term deployment.

2. Contributions

This Special Issue brings together ten cutting-edge studies in the field of metal research, covering shape memory alloys, magnetic functional materials, energy storage materials, and biomedical structural applications.

2.1. Shape Memory Alloys and Heusler Alloys

Shape memory alloys (SMAs) and their functional properties represent a primary focus of this Special Issue. Chen et al. (contribution 1) significantly enhanced the fatigue resistance of superelastic NiTi wires by combining mechanical training with nanocrystallization. Their study revealed that stress-controlled training induces a preferred grain orientation, which effectively reduces local stress concentrations and prolongs structural fatigue life. Timofeeva et al. (contribution 2) investigated the cyclic stability of superelasticity in quenched and aged NiFeGaCo single crystals. They highlighted the role of Ω-phase precipitates in cyclic stability, pointing out that degradation mechanisms involve the inhomogeneous distribution of dislocations around particles and the formation of residual martensite. Daróczi et al. (contribution 3) utilized acoustic emission (AE) techniques to provide significant insights into the microscopic mechanisms of rubber-like deformation in NiFeGaCo single crystals. They found that the asymmetry in AE activity observed during unloading is associated with the difficulty of renucleating stress-induced martensite variants.

Regarding magnetic Heusler alloys, Sakon et al. (contribution 4) performed magnetostriction measurements on the ferromagnetic Heusler alloy $Ni_2MnGaCu$. The study demonstrated that the alloy exhibits a large magnetostriction of up to 1300 ppm near the martensitic transformation temperature, confirming magnetic-field-induced strain during the transition from paramagnetic austenite to ferromagnetic martensite. Kamantsev et al. (contribution 5) focused on the Inverse Magnetocaloric Effect in NiMnSnCu Heusler alloys at low temperatures. Through direct measurements and theoretical calculations, they observed a significant adiabatic temperature change near the magnetic structural transition at 117 K under a 10 T field, offering new insights for cryogenic magnetic refrigeration. Furthermore, Liu et al. (contribution 6) innovatively designed a terahertz metamaterial absorber using Ni-Mn-Sn ferromagnetic shape memory alloy films. The device allows for dynamic switching between broadband and narrowband absorption via external magnetic fields, showcasing the potential of smart materials in terahertz applications.

2.2. *Magnetocaloric Materials*

Beyond Heusler alloys, rare-earth-based intermetallic compounds remain a key subject for magnetic refrigeration. Kuchin et al. (contribution 7) reported that the GdRuSi compound exhibits a giant magnetic entropy change and refrigerant capacity near the nitrogen liquefaction temperature (~77 K). Theoretical calculations confirmed that the magnetic moment primarily originates from Gd atoms, identifying GdRuSi as an ideal candidate material for nitrogen liquefaction magnetic refrigerators.

2.3. *Energy and Electronic Materials*

In the fields of energy storage and electronic packaging, this Special Issue features two studies with high practical value. Kwak et al. (contribution 8) synthesized MgH_2-$NaAlH_4$ composite hydrogen storage materials via mechanical ball milling. They found that the addition of 30 wt% $NaAlH_4$ significantly reduced the decomposition temperature of MgH_2 and increased the effective hydrogen storage capacity (up to 7.42 wt%), thereby improving the hydrogenation/dehydrogenation kinetics. Jung et al. (contribution 9) addressed the fabrication of electrodes for temperature-sensitive electronic devices, such as perovskite solar cells, by proposing a silver paste formulation based on azeotrope solvent application. This method achieved rapid sintering of Ag films at a low temperature of 140 °C, resulting in extremely low electrical resistivity and resolving thermal damage issues in flexible electronic manufacturing.

2.4. *Structural Materials for Medical Applications*

Addressing the issue of "stress shielding" in orthopedic implants, Mironovs et al. (contribution 10) explored methods to regulate the elastic modulus of 316L stainless steel plates through perforation design. Experimental and finite element simulation results

https://doi.org/10.3390/met16020153

showed that increasing the perforation area effectively lowers the material's Young modulus to more closely match that of human bone, thereby improving biomechanical compatibility.

3. Conclusions

We hope that this Special Issue of *Metals* not only reflects current scientific progress but also stimulates further collaboration across metallurgy, solid-state physics, mechanical design, and device engineering. We extend our sincere appreciation to all authors, reviewers, and editorial staff for their dedicated contributions. The advances showcased here reaffirm the central role of metallic functional materials in shaping future-adaptive, efficient, and high-performance technologies.

Conflicts of Interest: The authors declare no conflicts of interest.

List of Contributions:

1. Chen, P.; Cai, X.; Min, N.; Liu, Y.; Wang, Z.; Jin, M.; Jin, X. Enhanced Fatigue Resistance of Nanocrystalline $Ni_{50.8}Ti_{49.2}$ Wires by Mechanical Training. *Metals* **2023**, *13*, 361.
2. Timofeeva, E.; Panchenko, E.; Zherdeva, M.; Tokhmetova, A.; Chumlyakov, Y.I. The Cyclic Stability of the Superelasticity in Quenched and Aged $Ni_{44}Fe_{19}Ga_{27}Co_{10}$ Single Crystals. *Metals* **2023**, *13*, 1538.
3. Daróczi, L.; Kamel, S.M.; Tóth, L.Z.; Panchenko, E.Y.; Chumljakov, Y.I.; Beke, D.L. Acoustic Emission During Rubber-like Deformation in $Ni_{51}Fe_{18}Ga_{27}Co_4$ Single Crystalline Shape Memory Alloys. *Metals* **2025**, *15*, 131.
4. Sakon, T.; Morikawa, K.; Narumi, Y.; Hagiwara, M.; Kanomata, T.; Nojiri, H.; Adachi, Y. Magnetostriction of Heusler Ferromagnetic Alloy, $Ni_2MnGa_{0.88}Cu_{0.12}$, around Martensitic Transition Temperature. *Metals* **2023**, *13*, 1185.
5. Kamantsev, A.P.; Koshkid'ko, Y.S.; Gaifullin, R.Y.; Musabirov, I.I.; Koshelev, A.V.; Mashirov, A.V.; Sokolovskiy, V.V.; Buchelnikov, V.D.; Ćwik, J.; Shavrov, V.G. Inverse Magnetocaloric Effect in Heusler $Ni_{44.4}Mn_{36.2}Sn_{14.9}Cu_{4.5}$ Alloy at Low Temperatures. *Metals* **2023**, *13*, 1985.
6. Liu, R.; Wang, X.; Zhu, J.; Tian, X.; Zhao, W.; Tan, C.; Zhang, K. Terahertz Metamaterial Absorber Based on Ni-Mn-Sn Ferromagnetic Shape Memory Alloy Films. *Metals* **2023**, *13*, 1162.
7. Kuchin, A.G.; Platonov, S.P.; Mukhachev, R.D.; Lukoyanov, A.V.; Volegov, A.S.; Gaviko, V.S.; Yakovleva, M.Y. Large Magnetic Entropy Change in GdRuSi Optimal for Magnetocaloric Liquefaction of Nitrogen. *Metals* **2023**, *13*, 290.
8. Kwak, Y.-J.; Song, M.-Y.; Lee, K.-T. Improvement in the Hydrogen Storage Properties of MgH_2 by Adding $NaAlH_4$. *Metals* **2024**, *14*, 227.
9. Jung, S.H.; Park, J.E.; Lee, J.-H. Decreasing Electrical Resistivity of Ag Film by Low-Temperature Evaporation and Sintering through Azeotrope Application. *Metals* **2024**, *14*, 1123.
10. Mironovs, V.; Osipova, M.; Akishin, P.; Zemchenkovs, V.; Serdjuks, D. Methods for Evaluating the Elastic Properties of Stainless Steel Perforated Plates. *Metals* **2025**, *15*, 711.

https://doi.org/10.3390/met16020153

metals

MDPI

Article

Methods for Evaluating the Elastic Properties of Stainless Steel Perforated Plates

Viktors Mironovs [1], Marija Osipova [1], Pavel Akishin [2], Vjcheslavs Zemchenkovs [3] and Dmitrijs Serdjuks [4,*]

1 Institute of Mechanical and Biomedical Engineering, Riga Technical University, LV-1048 Riga, Latvia; viktors.mironovs@rtu.lv (V.M.); marija.osipova@edu.rtu.lv (M.O.)
2 Institute of High-Performance Materials and Structures, Riga Technical University, LV-1048 Riga, Latvia; pavels.akisins@rtu.lv
3 Institute of Materials and Surface Engineering, Riga Technical University, LV-1048 Riga, Latvia; vjaceslavs.zemcenkovs@rtu.lv
4 Institute of Civil Engineering, Riga Technical University, LV-1048 Riga, Latvia
* Correspondence: dmitrijs.serdjuks@rtu.lv

Abstract: Perforated materials are widely used in various fields, including in medicine, for example, in trays for placing and storing cutting tools and for sterilizing disposable materials. Currently, the effective elastic modulus of orthopedic plates is higher than the effective elastic modulus of human bone tissue (the effective elastic modulus of bone ranges between 10 and 30 GPa, depending on the type of bone). This difference in effective elastic modulus leads to the phenomenon known as the stress shielding effect, where the bone experiences insufficient mechanical loading. One potential approach to influence the effective elastic modulus of orthopedic plates is through perforations in their design. Stainless steel 316L has garnered significant interest among medical engineering specialists due to its lower weight, higher strength, and superior biocompatibility. The elastic properties of perforated constructions are influenced by their internal quality, dimensions, shapes, and the overall perforation area, making their study important. An experiment was conducted on perforated plates of 316L stainless steel with perforation areas ranging from 3% to 20%. Increasing the perforation area in perforated 316L stainless steel plates (perforated plates had dimensions of 50 mm in height, 20 mm in width, and 1 mm in thickness; hole diameter of 1 mm; and pitch between the holes of 2, 3, 4, and 5 mm) from 3% to 20% resulted in a decrease in Young's modulus of the perforated plates from 199 GPa to 147.8 GPa, determined using a non-destructive method for determining resonant frequencies using a laser vibrometer. A three-point bending test on the perforated plates confirmed these findings, demonstrating a consistent trend of decreasing Young's modulus with increasing perforation area, from 194.4 GPa at 3.14% to 142.6 GPa at 19.63%. The three-point bending method was also employed in this study to determine the Young's modulus of the perforated plates in order to reinforce the obtained results on the elastic properties by determining the resonance frequencies with a laser vibrometer. It was discovered that the Young's modulus of a perforated plate cannot be determined solely by the perforation area, as it depends on both the perforation diameter and the pitch between the perforations. In addition, finite element method (FEM) simulations were conducted, revealing that increasing perforation diameter and decreasing pitch significantly reduce the Young's modulus—with values dropping from 201.5 GPa to 72.6 GPa across various configurations.

Keywords: perforated metal plates; elastic properties; impulse excitation technique/methods; three-point bending; effective elastic modulus

1. Introduction

Metals such as stainless steel and titanium are extensively used in medical devices due to their corrosion resistance, durability, and biocompatibility. Stainless steel is commonly used for sterilization trays and surgical instrument storage, while titanium is favored for implants due to its strength and biocompatibility [1,2]. Perforation is the process of creating regular or irregular holes in a material [3,4]. One of the main advantages of perforated metal constructions is their lightness. Due to the holes, perforated metal constructions are considerably lighter than their solid counterparts, which reduces the overall weight of structures and facilitates transportation and installation (Figure 1).

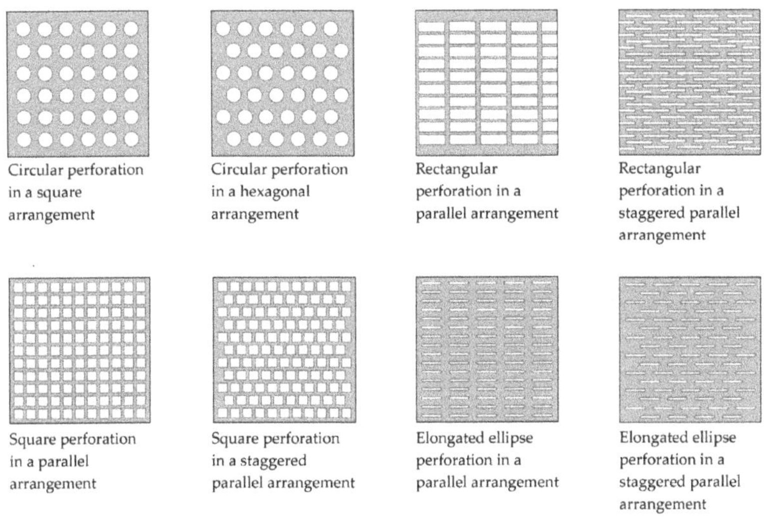

Figure 1. Types of perforation in metal construction.

Perforated constructions can be made from various substances, such as paper, cardboard, plastic, and metal. Metal perforation is carried out using methods such as stamping, laser cutting, waterjet cutting, and others. Stamping is a more efficient technology for mass production, providing a high processing speed and low cost for high volumes. However, stamping technology has disadvantages, such as high cost in small series production, as well as the difficulty of modifying the design of the structure. Also, stamping is not suitable for all materials and their thickness, and may also cause burrs and defects on the products, which makes it unacceptable for medical devices such as orthopedic implants [4]. Laser cutting technology provides the most precise and accurate perforation of structures; therefore, it is suitable for most materials and makes it possible to create shapes of various complexity [4]. Waterjet cutting is also suitable for all materials and provides high quality perforations, but without the heat of laser cutting. The perforation process with waterjet cutting is longer than that with laser cutting. The accuracy and precision of waterjet cutting is less than that of laser cutting, but better than that of punching [4].

Perforated materials can vary significantly depending on their application and technical requirements [5–8].

Perforated materials also offer sufficient strength and stiffness, making them suitable for constructions where both lightness and durability are important. Another important factor in the use of perforated materials is cost efficiency. The perforation process helps to save raw materials, reducing production costs [4,6,9]. Thus, perforated materials combine functional and economic advantages, making them highly sought after in various industrial and construction sectors [7,9,10]. Also, microperforated materials are a type of perforated metal material characterized by having very small holes, typically in the millimeter or sub-millimeter range. These materials are designed to provide specific properties that cannot be achieved with solid materials. The main features and applications of microperforated materials are highly effective for sound absorption without the need for additional classical absorbing materials, making them ideal for applications in machinery, aerospace, and industrial enclosures [11,12]. Despite their reduced weight, microperforated materials can still provide significant tensile strength and stiffness, depending on the source material [13]. They are used in construction for creating lightweight but strong structural elements. In medical applications, particularly in the field of biomaterial development and research, a key challenge is to optimize the physical and chemical properties of materials to enhance biocompatibility, reduce weight, and adjust elasticity to match that of bone while maintaining material strength.

Currently, perforation technology is being increasingly used for medical applications. For the production of sterilization trays and trays for washing instruments, it is most appropriate to use the waterjet cutting method for perforation, which provides higher productivity and economic efficiency. For the production of implants, the laser perforation method is preferred, which provides holes in plates and structures of small diameter [14–16]. These applications include perforation technology in the construction sector.

Implants used in medicine, such as orthopedic implants, dental implants, and bone fixators, are subject to stress shielding, which occurs when an implant bears the mechanical loads that would otherwise be carried by the bone. This happens because the implant, often made from a material stiffer than the surrounding bone, takes on a disproportionate share of the load, reducing the stress on the bone [17]. Stress shielding can lead to a reduction in bone density and strength because the bone is not subjected to the usual physiological stresses that stimulate remodeling and maintenance [17]. This can lead to a reduction in the mechanical stress experienced by the bone, which in turn can cause bone resorption and weakening over time. In the context of biomedical implants, stress shielding is a critical issue because it can compromise the long-term success of the implant. For example, metal implants, which are much stiffer than bone, can lead to significant stress shielding. This mismatch in stiffness results in the bone not being sufficiently loaded, which is necessary for maintaining bone density and strength [18]. To prevent mechanical stress shielding—where implants with high elastic properties take on most of the load, leading to the atrophy and decreased density of the surrounding bone tissue known as osteopenia or osteoporosis—it is important to reduce elastic properties such as the Young's modulus of construction for orthopedic implants [17]. Constructions with a Young's modulus close to the elastic properties of human bone (Young's modulus of human bone is 10–30 GPa) provide better load distribution between the orthopedic implant and the bone, reducing the risk of bone degradation [19,20].

The primary objective of this article is to determine whether perforations in metal could be used to affect the construction's elastic properties to an optimal level for orthopedic implants. By varying the percentage of the perforated area, along with controlling the shape, size, and pattern of the holes, it is possible to adjust the Young's modulus of the implant construction to match that of the specific bone into which the orthopedic implant will be inserted.

According to Hooke's law, for an isotropic body:

$$\sigma = E\epsilon \tag{1}$$

where σ is the stress, depending on force and cross-sectional area, the Young's modulus of construction is related to the cross-sectional area of the construction under stress [21,22].

Specifically, Young's modulus is determined by the ratio of applied stress to strain, where stress is the force per unit area [23]. As the cross-sectional area of construction affects the distribution of the applied force, any change in will directly influence the magnitude of stress, and, therefore, the calculation of E [24]. Thus, for a given construction, variations in the cross-sectional area (depending on perforation diameter and pitch between perforations) can significantly impact the construction's apparent stiffness and deformation response under load. The integration of perforations into the metal structure, specifically in this research into a metal plate, reduces the cross-sectional area. Consequently, the method of perforation influences the Young's modulus of construction, effectively lowering it.

The three-point bending test is a standardized destructive method widely used in materials science to assess the mechanical properties of materials, particularly their flexural behavior and Young's modulus [25]. This method provides critical information on the response of materials to bending loads, which is particularly relevant in the context of orthopedic implant manufacturing, where components frequently experience flexural stresses during service. In this method, a rectangular or beam-like specimen is positioned on two parallel supports, and a concentrated load is applied at its center. As the load increases, the specimen bends, and the corresponding force–displacement response is recorded. The calculation of Young's modulus, based on the Euler–Bernoulli beam theory, uses the following equation:

$$E = \frac{F\,L^3}{4dbh^3} \tag{2}$$

where E is the Young's modulus, F is the applied force, L is the span length between supports, b and h are the width and thickness of the tested specimen, and d is the deflection at the load point [25].

Finite element method (FEM) analysis is widely used to investigate the mechanical behavior of perforated materials under various loading conditions. The presence of perforations significantly influences the stress–strain distribution, load capacity, and energy absorption characteristics of materials, which are crucial for their application in engineering structures. These analyses provide valuable insights for optimizing design parameters to enhance the structural efficiency of perforated material [26]. The elastic properties of materials are currently mostly determined using mechanical compression or tension tests, in which the deformation of the material under load is measured. In addition to these methods, bending, torsion, and dynamic loading tests are used to evaluate the behavior of the material under different conditions [4].

Summarizing the relevance of perforated materials and the necessity to optimize the elastic properties of constructions for orthopedic implants, the primary objective of this article was to investigate the effect of varying perforation areas and hole diameter and pitch between holes on the elastic properties of these constructions, and to explore the potential application of using this technology in orthopedic implant design.

2. Materials and Methods

2.1. Stainless Steel 316L Plate Perforation

The main task of this article is to analyze the dependence of the elastic properties of 316L stainless steel perforated plates on the perforation area. For the experiment, austenitic stainless steel AISI 316L was selected due to its low carbon content. It is well known for its high corrosion resistance, especially in aggressive environments and against intergranular corrosion. AISI 316L contains approximately 16–18% Cr, 10–14% Ni, and up to 2% Mo, which provides this material with greater mechanical strength compared to other stainless steel grades. However, other grades can also be used—for example, AISI 304, which offers increased ductility, a property that is essential for the fabrication of complex-shaped cranial implants. The stainless steel 316L metal alloy was selected for the experiment due to its lower weight, greater durability, and higher biocompatibility [27–30]. In this study, only one specimen was tested for each perforation percentage. The primary objective of this study was to explore the conceptual potential of perforation to modulate the elastic properties of orthopedic implants, with less emphasis placed on deriving statistically reliable or standardized values for each specific perforation configuration. Specifically, this study highlights how perforation can effectively reduce the Young's modulus of metallic implant materials, thereby offering a prospective approach to reduce the stress shielding effect. Four stainless steel 316L plates with dimensions of 50 mm in height, 20 mm in width, and 1 mm in thickness were manufactured and fabricated using the same equipment, materials, and parameters to ensure consistency across tests. The perforations in all specimens were created individually on solid stainless steel 316L plates using laser cutting, rather than cutting shapes from pre-perforated sheets. This ensured full control over the perforation geometry and positioning. The holes were arranged in a regular square pattern and oriented symmetrically, with the perforation grid aligned perpendicular to all plate edges. Special care was taken to preserve exact distances between perforations and from the perforations to the edges of the specimen. The precision of the cutting process was verified, confirming that the perforations were consistent and accurately positioned relative to the bending direction. Laser cutting was used for the stainless steel 316L plates, and the perforation parameters are shown in Table 1. The determination of resonance frequencies using a laser vibrometer Polytec PSV-500 (Polytec GmbH, Waldbronn, Germany) was employed to determine the Young's modulus of the perforated stainless steel 316L plates. The results from the laser vibrometry were processed to determine the dependence of the Young's modulus of the perforated stainless steel 316L plates on the perforation area. The perforation type was circular in a square arrangement. The perforation area was calculated using the following formula:

$$Aperf = \frac{d^2 78.54}{s^2} \tag{3}$$

where d is the hole diameter and s is the pitch between holes [4].

Table 1. Perforation parameters of the stainless steel 316L plates.

Sample No.	Perforation Diameter (mm)	Pitch between Perforations (mm)	Perforation Area (%)
1.	1	5	3.14
2.	1	4	4.91
3.	1	3	8.78
4.	1	2	19.63

Figure 2 shows the samples of perforated stainless steel 316L plates created using the laser cutting method, with perforation areas ranging from 3.14% to 19.63%.

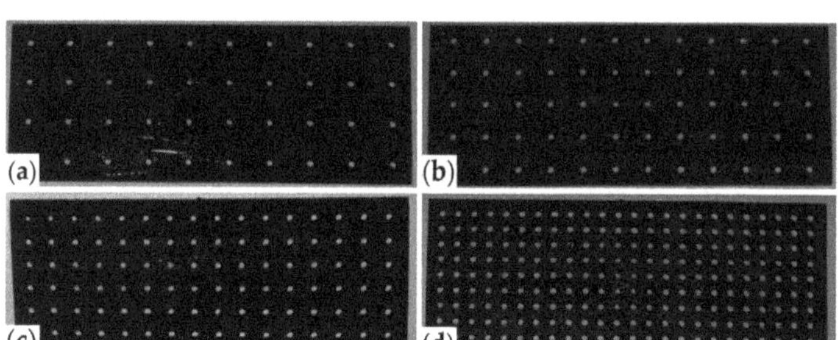

Figure 2. Perforated stainless steel 316L plates made using the laser cutting method. (**a**) A_{perf} = 3.14%; (**b**) A_{perf} = 4.91%; (**c**) A_{perf} = 8.78%; (**d**) A_{perf} = 19.63%.

2.2. Perforated Stainless Steel 316L Plates' Elastic Properties Determination Using Laser Vibrometry

One of the achievements of this article was determining the elastic properties of perforated metal plates using a non-destructive method. Laser vibrometry is an instrument for determining the dynamic properties of the sample—the resonant frequencies, which in turn are needed to determine the elastic properties. The resonant methods widely used for the determination of elastic properties of composite materials were adopted for the characterization of the perforated steel plates in this study [31–36]. A laser vibrometer was used as a tool for the non-contact experimental determination of resonant frequencies. Laser scanning vibrometry is based on detecting the Doppler frequency shift that occurs when laser light is scattered from a moving surface [37]. This frequency shift is directly proportional to the surface velocity, allowing for the precise measurement of vibration speed. The advantages of this method include non-contact measurements that do not affect the structure of material, high resolution, and accuracy [37,38]. Laser vibrometry is particularly suitable for analyzing the elastic properties of perforated metal plates because it can precisely measure vibration speed and patterns without mechanically influencing the material. The main components of the laser vibrometry system include a laser source, which generates a laser beam directed at the surface of the sample under examination [37]. To determine the elastic properties, including the Young's modulus, of the perforated 316L stainless steel plates with perforation areas ranging from 3.14% to 19.63%, the Polytec PSV-500 system was used (Figure 3). All measurements were conducted under environmental conditions, where room temperature was maintained at approximately 19 °C, with a relative humidity of 45%. Each specimen was tested once.

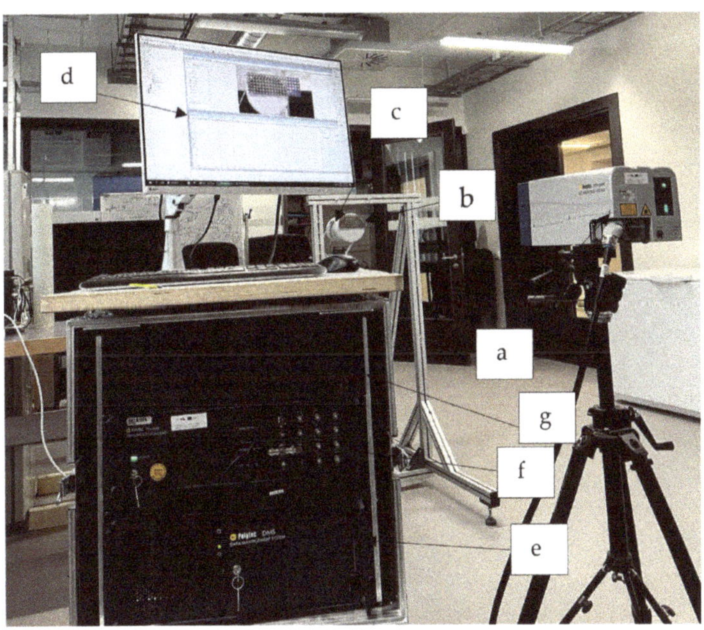

Figure 3. The Polytec PSV-500 scanning head was used in this experiment to determine the resonance frequences using the laser vibrometry method. (**a**) Scanning head; (**b**) sample; (**c**) loudspeaker; (**d**) monitor; (**e**) PC; (**f**) junction box; (**g**) controller.

In the control software, the excitation frequency range was set from 0 to 10 kHz, covering the resonance frequencies of the plates. Each plate's surface was scanned, evenly distributed across the length and width, to obtain a detailed view of the vibrations. In Figure 4, the process of scanning the surface of the plate is shown at points that were set in the software before starting scanning. The scanning points were uniformly distributed across the plate surface with a spacing of 2–3 mm between adjacent points.

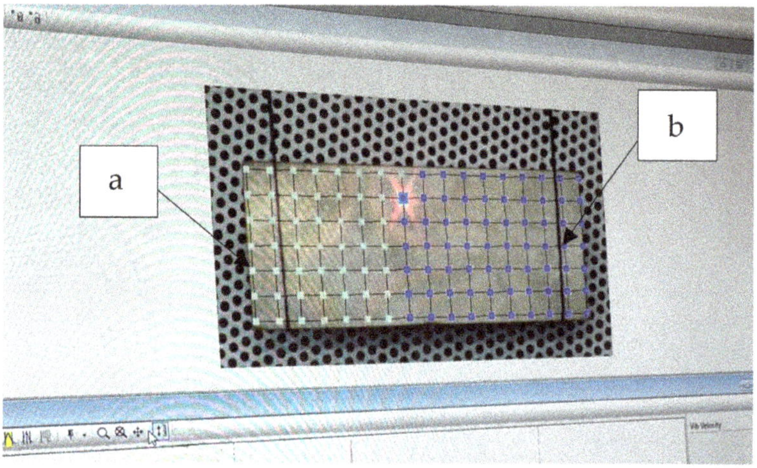

Figure 4. Sample scanning process using Polytec PSV-500. (**a**) Scanned points marked with light green color; (**b**) un-scanned points marked with blue color.

The data were transmitted to a computer, where they were processed to generate frequency response graphs. The 2.0 mode label, which is shown in Figure 5 (two half-waves in length and zero in width) indicates the fundamental bending frequency, where the material experiences maximum stress at the center and minimum at the edges. The 2.0 mode label represents the vibration modes used to calculate the Young's modulus of the perforated plates [36].

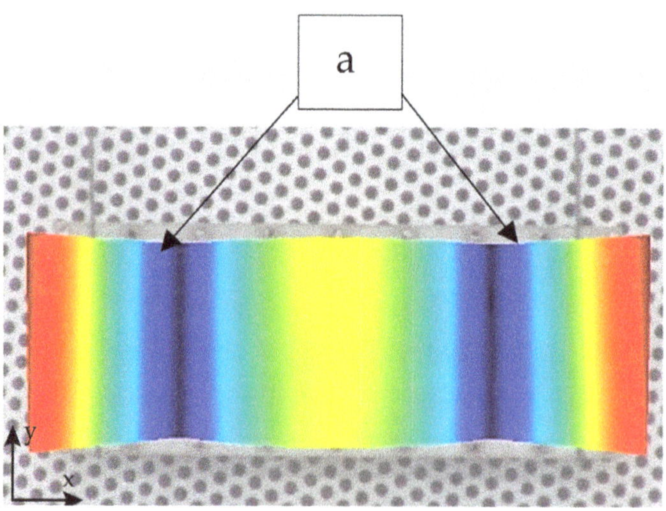

Figure 5. Stainless steel 316L 4 sample's experimentally measured mode shape 2.0 as the fundamental bending frequency (f_b). (a) Two half-waves in length.

The measured values of the resonance frequencies were applied to the appropriate formulas to calculate the Young's modulus of the plates. Each perforated stainless steel 316L sample was tested using the laser vibrometry method. After each test, the results included the fundamental bending frequency. Table 2 shows the obtained data on the fundamental bending frequencies for each tested sample.

Table 2. Physical, geometric, experimental, and calculated data of the perforated stainless steel 316L samples.

Sample No.	Weight (g)	Length (mm)	Width (mm)	Thickness (mm)	Fundamental Bending Frequency (kHz)	Correction Factor T	Perforated Plates' Young's Modulus (GPa)
1.	7.4	50	20	1	2.100	1.003	193.63
2.	7.4	50	20	1	2.077	1.003	189.41
3.	7.2	50	20	1	2.048	1.003	179.18
4.	6.6	50	20	1	1.943	1.003	147.84

The formula for calculating Young's modulus E of the tested samples using the sample's fundamental bending frequency f_b is as follows:

$$E = 0.9465 \times m \times \frac{f_b^2}{b} \times \left(\frac{l}{h}\right)^3 \times T \quad (4)$$

where geometrical parameters for calculating Young's modulus of tested samples were thickness h, length l, and width b, and the physical parameter was weight m. Correction factor T (which for $l/h > 20$) was calculated as follows:

$$T = 1 + 6.585\left(\frac{h}{l}\right)^2 \quad (5)$$

Using experimentally obtained fundamental bending frequencies, Young's modulus E of the tested samples was calculated according to Formulas (4) and (5) [34,36]. To perform the calculations, geometric parameters such as length, width, and thickness were measured using a caliper (accuracy of ±0.1 mm), and the weight of the samples was weighed (accuracy of ±0.1 g). The obtained data for the calculations of the Young's modulus of perforated stainless steel 316L plates are shown in Table 2.

2.3. *Perforated Stainless Steel 316L Plates' Elastic Properties Determination Using Three-Point Flexural Test*

The three-point bending test was conducted to determine the elastic properties of four perforated stainless steel 316L plates. The perforation area varied between 3.14% and 19.63%, and the experiment followed a destructive testing approach. All measurements were conducted under environmental conditions, where room temperature was maintained at approximately 21 °C, with a relative humidity of 45%. Each specimen was tested only once, due to the fact that a single sample was available for each perforation configuration. Additionally, the three-point bending test employed in this study is a destructive method, rendering the specimen unusable for repeated testing. This approach, while limiting statistical analysis, was considered sufficient for the conceptual validation of the proposed methodology. The tests were carried out using an ElectroPlus E3000 testing machine (Instron, Norwood, MA, USA) (Figure 6). The support span (distance between the two outer supports) was set to 41 mm. Each specimen was subjected to loading, and numerical data were recorded, including the applied force F and flexure extension. The experimental data were subsequently analyzed, and the Young's modulus of each perforated plate was calculated using the Euler–Bernoulli beam theory and Equation (2) in MS Excel.

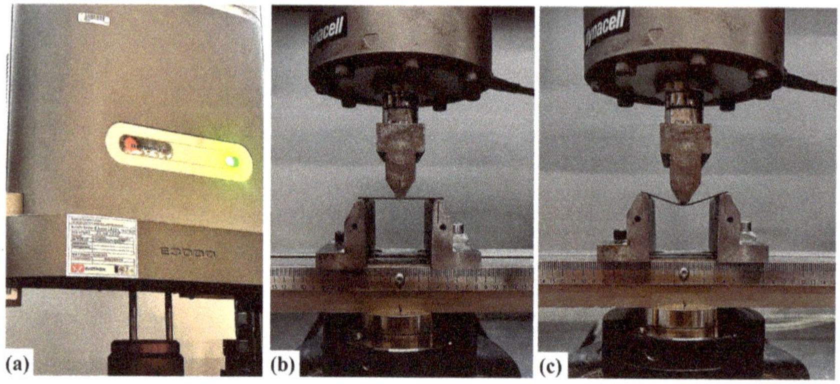

Figure 6. Three-Point bending test of the perforated stainless steel 316L plates. (**a**) Experimental machine for conducting the three-point bending test; (**b**) perforated stainless steel 316L plate before the three-point bending test; (**c**) perforated stainless steel 316L plate after the three-point bending test.

2.4. Perforated Stainless Steel 316L Plates' Elastic Properties Determination Using the Finite Element Method (FEM)

A virtual experiment using the finite element method was used to investigate the elastic properties of 316L stainless steel. The main objective of this experiment was to study the impact of perforation parameters, such as hole diameter and pitch between holes, on the Young's modulus of perforated 316L stainless steel plates. These parameters determine the total perforation area, which allows for a more detailed investigation of their effect on the elastic properties of the plates. This experiment was conducted using the finite element method (FEM) to obtain new results on the dependence of Young's modulus of perforated stainless steel 316L plates on the perforation configuration.

To enhance the range of variable parameters and systematically evaluate the effect of the perforation diameter and pitch between perforations on the Young's modulus of steel plates, the experimental framework was broadened. The investigation focused on perforation diameters of 1.5, 2.0, 2.5, and 3.0 mm, combined with consistent pitches of 2, 3, 4, and 5 mm between the holes. For the experiment, the dimensions of the 316L stainless steel plates were created according to the dimensions of the original plates, with a length of 50 mm, width of 20 mm, and thickness of 1 mm. To optimize research efficiency and reduce time expenditure, a numerical modal analysis was employed as a substitute for physical testing. Numerical simulations of the perforated plates were performed using ANSYS Mechanical APDL 2023R1, where structural finite elements (Shell181) were used to accurately model the geometry and mechanical behavior of the plates (Figure 7a). Undamped modal analysis was executed using the Block Lanczos method (Figure 7b). This approach allowed for a detailed assessment of how variations in perforation parameters influence the elastic characteristics.

The elastic and mass properties of 316L stainless steel, required for the numerical experiments, were determined by minimizing the differences between the physical and numerical mass and the first bending eigenfrequency of plates with a perforation diameter of 1.0 mm and pitches between perforations of 2, 3, 4, and 5 mm. The identified properties are as follows: density $\rho = 7800$ kg/m^3, Young's modulus $E = 215$ GPa, and Poisson's ratio as 0.3. Notably, the identified Young's modulus E is higher than the typical datasheet range of 193–200 GPa. However, the tensile tests have also yielded values within the range of 210–215 GPa, corroborating the results obtained in this study [38]. Table 3 presents the results and the difference in physical and numerical mass and dynamic properties of the 316L stainless steel plates with a perforation diameter of 1 mm.

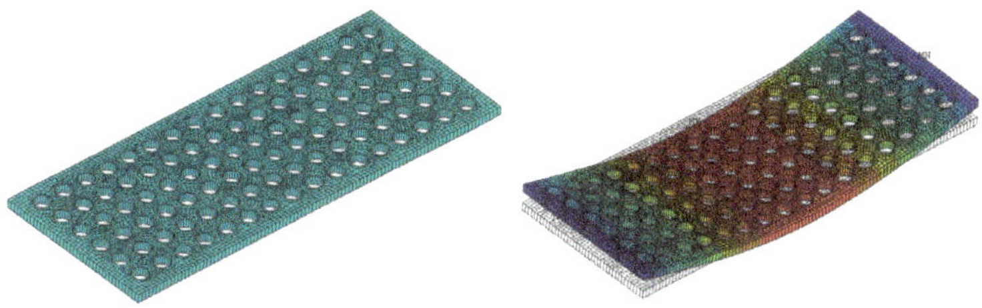

(**a**) (**b**)

Figure 7. Finite element model of the stainless steel 316L perforated plate with a perforation diameter of 2 mm and pitch of 3 mm. (**a**) Finite element model; (**b**) 1st bending mode shape.

Table 3. Comparison of the physical and numerical properties of the perforated stainless steel 316L plates with a perforation diameter of 1.0 mm.

Pitch Between Holes (mm)	Experimental 1st Eigenfrequency (kHz)	Numerical 1st Eigenfrequency (kHz)	1st Eigenfrequency Difference (%)	Experimental Weight (g)	Numerical Weight (g)	Weight Difference (%)
5	2.100	2.121	1.0	7.40	7.55	2.0
4	2.077	2.089	0.6	7.40	7.43	0.4
3	2.048	2.047	0.0	7.20	7.21	0.1
2	1.943	1.922	−1.1	6.60	6.48	−1.8

The numerically obtained data for calculating the Young's modulus of the perforated 316L stainless steel virtual plates using Formulas 4 and 5 are summarized in Table 4.

Table 4. Physical, numerical, and calculated data of the perforated stainless steel 316L virtual samples with dimensions of 50 × 20 × 1 mm.

Sample No.	Perforation Diameter (mm)	Pitch Between Perforations (mm)	Weight (g)	Fundamental Bending Frequency (kHz)	Correction Factor T	Young's modulus of Perforated Virtual Samples (GPa)	Perforation Area (%)
V1	1	5	7.55	2.121	1.003	201.53	3.14
V2	1	4	7.43	2.089	1.003	192.38	4.91
V3	1	3	7.21	2.047	1.003	179.26	8.78
V4	1	2	6.48	1.922	1.003	142.03	19.63
V5	1.5	5	7.25	2.076	1.003	185.39	7.07
V6	1.5	4	6.97	2.010	1.003	167.08	11.05
V7	1.5	3	6.48	1.934	1.003	143.81	19.64
V8	1.5	2	4.82	1.729	1.003	91.84	44.18
V9	2	5	6.82	2.018	1.003	164.79	12.57
V10	2	4	6.33	1.910	1.003	137.02	19.64
V11	2	3	5.45	1.806	1.003	105.47	34.91
V12	2.5	5	6.27	1.950	1.003	141.46	19.64
V13	2.5	4	5.50	1.792	1.003	104.79	30.68
V14	2.5	3	4.12	1.658	1.003	67.19	54.54
V15	3	5	5.59	1.873	1.003	116.36	28.27
V16	3	4	4.49	1.651	1.003	72.62	44.18

In this experiment, the number of perforated plates is not consistent across samples with perforation diameters of 2, 2.5, and 3 mm, which is shown in Table 4. This difference is due to the difference between the increasing perforation diameter and the corresponding pitch between perforations. At certain pitch values, perforations could not be fitted within the plate dimensions without overlapping or breaching the required spacing between them, resulting in the absence of results for those configurations.

3. Results

3.1. Elastic Properties of Perforated Stainless Steel 316L Plates Obtained by Method of Determination of Resonant Frequencies and by Three-Point Flexural Test

Numerical data on flexure load and flexure extension were obtained for each perforated stainless steel 316L plate specimen.

The plot of the dependence of Young's modulus of perforated 316L stainless steel plates on perforation obtained by the method of determination of resonant frequencies and three-point flexural test in Figure 8 shows that, as the perforation percentage increases, the material's elastic properties decrease. The perforation area affects the Young's modulus of the structure, and, by this method, the Young's modulus of the structure can be achieved to solve the stress shielding effect. As the perforation area is 3.14%, the Young's modulus of the perforated plate is 193.63 GPa (by the method of determination of resonant frequencies) and 194.432 GPa (by the three-point flexural test). With an increase in perforation area to 4.91%, the decreasing trend continues, and the Young's modulus of the perforated plate reaches 189.41 GPa (by the method of determination of resonant frequencies) and 179.839 GPa (by the three-point flexural test). When the perforation area increases to 8.78%, the Young's modulus of the perforated plate decreases further to 179.18 GPa (by the method of determination of resonant frequencies) and 174.113 GPa (by the three-point flexural test). The most significant reduction in all parameters occurs at a perforation area of 19.625%, where the Young's modulus of the perforated plate falls to 147.84 GPa (by the method of determination of resonant frequencies) and 142.568 GPa (by the three-point flexural test). Summarizing the Young's modulus values, in the perforation area range from 3.14% to 8.725%, there is a gradual decrease in Young's modulus (about a 20 GPa difference). However, in the perforation area range from 8.725% to 19.625%, a more substantial decrease in Young's modulus is observed (about 30 GPa difference). We must consider the Young's modulus of human bone, which ranges from 10 to 30 GPa and is indicated in green in the plot of the Young's modulus of the perforated plates' dependence on the perforation area for perforated 316L stainless steel plates. The sample with the highest perforation area is 19.625%, which is still higher than the Young's modulus of human bone. Therefore, further research into 316L stainless steel plates with perforation areas greater than 19.625% would be necessary to reduce the Young's modulus to be closer to that of human bone.

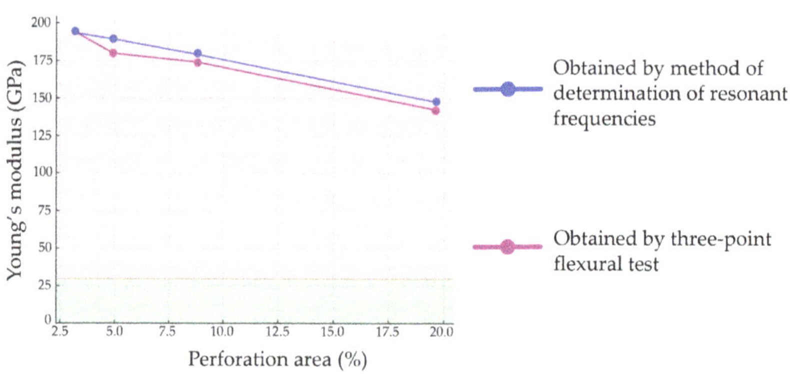

Figure 8. Dependence of Young's modulus of the perforated 316L stainless steel plates on perforation area (perforated plates dimensions of 50 mm in height, 20 mm in width, and 1 mm in thickness; hole diameter of 1 mm; and pitch between the holes of 2, 3, 4, and 5 mm), obtained by method of determination of resonant frequencies and obtained by three-point flexural test.

A static analysis of the experimental results was performed to provide an approximation of the perforated plates' elastic properties' dependence on the perforations' geometrical configuration. This is fundamental for reliably predicting material behavior under various perforation design configurations, especially for orthopedic implant applications, where elastic modulus control is necessary to mitigate the stress shielding effect.

The analysis was conducted using the EDAOpt v.2.96 software package, which enables the modeling of experimental data using polynomial regression and the evaluation of statistical reliability [39]. The independent variable was defined as the distance between the centers of the perforation holes (pitch), with tested values of 5 mm, 4 mm, 3 mm, and 2 mm, corresponding to perforation areas of 3.14%, 4.91%, 8.78%, and 19.63%, respectively. The dependent variable was the experimentally determined Young's modulus, derived from resonance frequency measurements using laser vibrometry. The results of the polynomial approximation, obtained using the EDAOpt v.2.96 software, are presented in Table 5.

Table 5. Results of the second-order polynomial regression.

Free Term	Linear Term	Quadratic Term	σ_{cross}, %	$R^2_{adjusted}$	σ, %	Standard Deviation of Expression	Maximal Relative Error, %
51.275	62.22	−6.78	54.39	0.97	16.32	20.69	1.27

The data were modelled using a second-order polynomial regression model, resulting in the following expression:

$$E = 51.275 + 62.22 \times P - 6.78 \times P^2 \quad (6)$$

where P is the pitch between perforations (in mm) and E is the predicted Young's modulus. The model's statistical parameters demonstrate the adjusted coefficient of determination, $R^2 = 0.97$, indicating that approximately 97.3% of the variance in the Young's modulus is explained by the model.

The significance analysis of normalized coefficients revealed that the linear term X has the dominant influence on the response, followed by the quadratic term X2, which introduces the necessary curvature to accurately model the observed non-linear relationship. The free term is constant and was not shown in the normalized coefficient analysis, as it does not contribute to the relative variation. The obtained regression clearly confirms that a decrease in pitch (leading to an increase in perforation area) results in a systematic reduction in Young's modulus. Importantly, the presence of a significant quadratic term indicates that this relationship is not strictly linear. The reduction in modulus accelerates at smaller pitch values, which is consistent with the physical understanding of stress concentration effects and loss of cross-sectional integrity. The model allows reliable estimation within the tested range of perforation pitches and provides a basis for predicting the elastic behavior of new designs with similar geometric parameters. Furthermore, it supports the experimental conclusion that both pitch and hole diameter must be jointly considered to accurately predict the mechanical response, as identical perforation areas with different geometries yield different Young's modulus values. The static analysis validates and reinforces the experimental findings by providing a statistically robust approximation of the relationship between perforation geometry and elastic properties. The resulting regression model can serve as a practical tool for pre-selecting perforation configurations in the design of orthopedic plates and other medical devices, with the objective of tuning elastic properties to approach the target modulus of human bone tissue and minimize the risk of stress shielding.

3.2. Elastic Properties of Perforated Stainless Steel 316L Plates with Variable Perforation Diameter and Pitch Between Perforations Using the Finite Element Method

When analyzing the results of the virtual experiment, we explored how variations of perforation diameter and pitch between perforations influence the elastic properties of perforated 316L stainless steel plates. The results were further visualized graphically as a 3D surface, demonstrating the relationship between the Young's modulus of the perforated plates, perforation diameter, and pitch between perforations. The analysis of the results demonstrates how an increase in perforation diameter and pitch between perforations leads to a decrease in the elastic modulus of the plates. A perforated plate with a larger perforation diameter and smaller pitch between the perforations consequently allows the Young's modulus to be reduced more rapidly than one with a smaller perforation diameter and larger pitch between the perforations, as shown in the 2D plot in Figure 9.

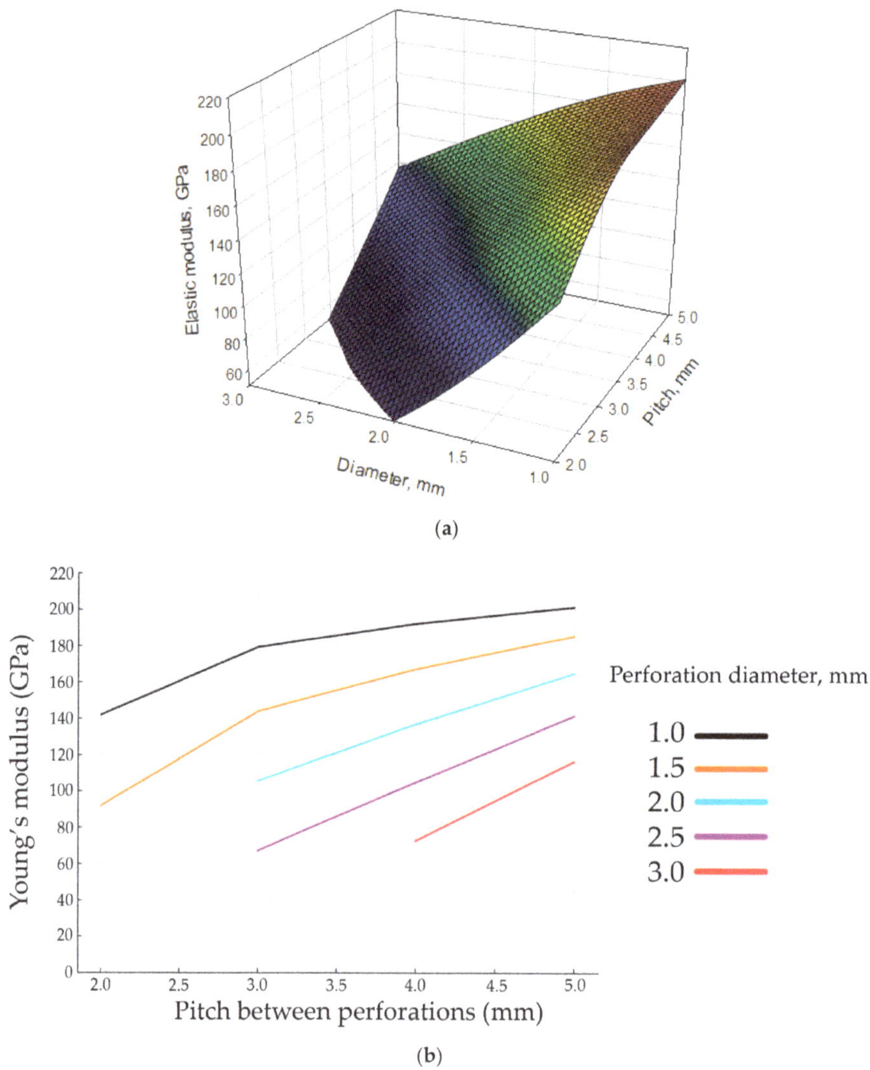

Figure 9. Dependence of the perforated stainless steel 316L plate's Young's modulus on perforation diameter and pitch between perforations. (**a**) Three-dimensional surface of elastic modulus, diameter, and pitch dependence; (**b**) 2D plot of Young's modulus dependence on pitch between the perforations and perforation diameter from 1mm to 3 mm.

A detailed statistical analysis of the virtual experiment results was performed using the EDAOpt v.2.96 software package to quantify and generalize the influence of perforation geometry on the elastic properties of perforated plates. The aim was to develop a regression model that could predict the effective Young's modulus E of perforated stainless steel plates based on two key parameters—the pitch between perforations and the perforation diameter. This approach allows the perforation pattern to be optimized in order to achieve the desired mechanical properties, which is particularly important for biomedical applications. A second-order polynomial regression model was selected as the optimal compromise between accuracy and physical interpretability. This dependence was approximated using a second-order polynomial function:

$$E = 144.767 + 52.3702 \times P - 109.531 \times D - 7.07648 \times P^2 + 17.2956 \times P \times D - 5.0348 \times D^2 \quad (7)$$

where E is the effective Young's modulus in GPa, P is the pitch between perforations in mm, and D is the perforation diameter in mm. The regression coefficients were calculated based on sixteen experimental points (virtual samples).

Table 6 shows the statistical quality of the regression model, with a key metric being the adjusted coefficient of determination, $R^2_{adj} = 0.992$, which indicates a model fit of 99.2% (i.e., 99.2% of the variance is explained).

Table 6. Results of the second-order polynomial regression for the finite element method.

Free Term	Linear Terms	Quadratic Terms	σ_{cross}, %	$R^2_{adjusted}$	σ, %	Standard Deviation of Expression	Maximal Relative Error, %
144.767	52.3702 −109.531	−7.07648 −5.0348 17.2956	11.65	0.99	8.79	42.12	7.06

Figure 10 presents the relative influence of the regression terms. Both primary variables (P and D) significantly affect the Young's modulus.

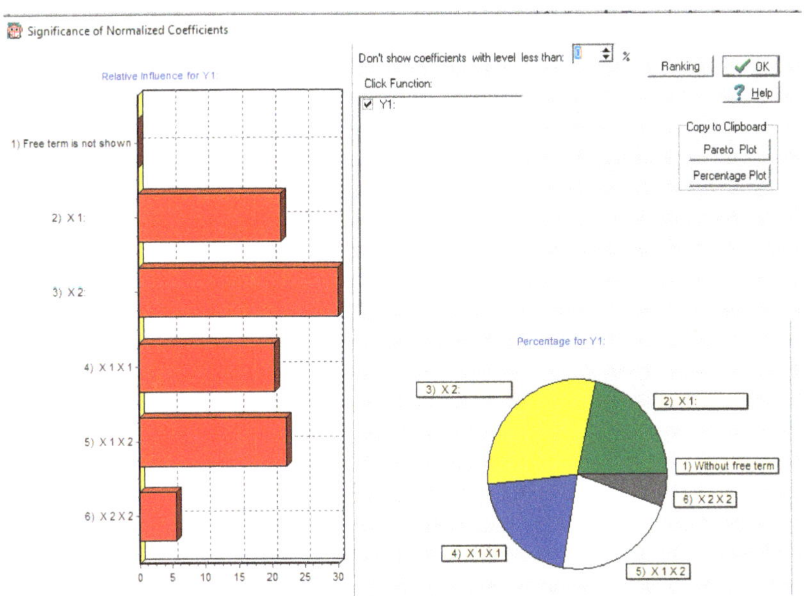

Figure 10. Relative influence of P and D parameters.

The largest influence is attributed to perforation diameter $D = X_2$ and pitch $P = X_1$. The interaction term $P*D$ also makes a strong contribution, and the quadratic terms P^2 and

D^2 refine the model curvature. Thus, it is confirmed that Young's modulus depends not only on the perforation area, but also on the specific combination of diameter and pitch.

To verify the model, three additional virtual samples with geometrical configurations that were not used during its construction were tested (Table 7), and we computed their values using an approximation model.

Table 7. Verification with additional points.

Sample No.	Perforation Diameter (mm)	Pitch Between Perforations (mm)	Weight (g)	Fundamental Bending Frequency (kHz)	Correction Factor T	Young's Modulus of Perforated Virtual Samples (GPa)	Young's Modulus of Perforated Virtual samples Obtained Using Mathematical Model (GPa)	Difference, %	Perforation Area (%)
T1	1.75	3.5	6.49	1.960	1.003	147.93	140.21	5.21	19.64
T2	2.25	2.5	3.46	1.631	1.003	54.61	56.82	4.05	63.62
T3	2.75	4.5	5.76	1.895	1.003	122.73	111.88	8.85	29.33

These samples provide an independent test of the model's predictive ability. The predicted Young's modulus values for these additional points showed good agreement with the calculated results (difference of 5.21% and 4.05%), confirming the sustainability of the approximation (difference below 10% is sustainable in our conditions) and its applicability to a wider design space.

In addition, a Pearson correlation analysis was conducted to compare the Young's modulus values obtained using three different methods: the determination of resonant frequencies, the finite element method, and the three-point bending test. The Pearson correlation coefficients are shown in Table 8.

Table 8. Pearson correlation coefficients between methods.

Methods	Pearson Correlation Coefficient
Determination of resonant frequencies and three-point bending test	0.98
Determination of resonant frequencies and finite element method	0.99
Finite element method and three-point bending test	0.99

These high correlation values confirm the consistency and reliability of the model and methodology even further. The graphical 3D surface representation offers a demonstrative view of this dependency. It was also discovered, based on the results shown in Table 4, that when the perforation areas are the same or close to each other, the Young's modulus of the perforated plates can be different. For perforated plates with a perforation area of 19.64% but different perforation diameters and pitch between the perforations, the Young's modulus of the perforated plates is different. When the perforation area is 19.64%, the perforated plate with 1 mm of perforation diameter and 2 mm pitch between perforations has a Young's modulus of 142.03 GPa; the perforated plate with 1.5 mm of perforation diameter and 3 mm pitch between perforations has a Young's modulus of 143.81 GPa; and the perforated plate with 2 mm of perforation diameter and 4 mm pitch between perforations has a Young's modulus of 137.02 GPa. The Young's modulus of perforated plates with a perforation area of 44.18% is also different for perforated plates with changing parameters, such as perforation diameter and pitch between perforations. At a perforation area of 44.18%, a perforated plate with a perforation diameter of 1.5 mm and a pitch between perforations of 2 mm has a Young's modulus of 91.84 GPa; and a perforated plate with a perforation diameter of 3 mm and a pitch between perforations of 4 mm has a Young's modulus of 72.62 GPa.

4. Discussion

4.1. Methods for Reducing the Elastic Modulus

In recent years, various approaches have been explored to reduce the elastic modulus of metallic implants and mitigate stress shielding. One promising approach is the development of β-type titanium alloys with an inherently low Young's modulus (below 60 GPa). Ti-29Nb-13Ta-4.6Zr (TNTZ) alloys and Ti-Nb-Zr-Sn systems, for example, achieve a combination of low stiffness and high fatigue strength through severe plastic deformation, thermal treatments, and the addition of non-toxic elements [40]. Another widely studied method involves fabricating porous titanium foams with interconnected porosity [41]. Their electrochemical dealloying technique produces foams with elastic moduli in the range of 15.5–36 GPa, approaching that of cortical bone, while maintaining sufficient mechanical strength for load-bearing applications. In comparison, the present study introduces a more controlled geometric approach involving the adjustment of the pitch and diameter of perforations in stainless steel plates. Although the absolute reduction in Young's modulus achieved with perforations (down to approximately 110–120 GPa) does not yet reach the levels obtained with specialized titanium alloys or foams, this method offers several practical advantages. It can be implemented using conventional manufacturing technologies, such as laser cutting. It preserves the structural integrity of the bulk material and allows the modulus to be tailored without introducing material heterogeneity. Furthermore, optimizing perforation patterns could enable further modulus reduction, bringing the mechanical behavior of implants closer to that of bone without the need for more complex alloy design or porosity processing.

4.2. The Method of Determination of Resonant Frequencies, the Three-Point Flexural Test Method, and the Finite Element Method Obtained Results

The calculated percentage reductions are summarized in Table 9, providing a comparative assessment of the different methods. The results show the importance of selecting an appropriate evaluation technique when analyzing the mechanical properties of perforated constructions, as different methods may yield varying magnitudes of stiffness reduction. A crucial aspect of evaluating the reliability of the obtained Young's modulus values is assessing the consistency of the results across different methods. If the percentage difference between methods varies significantly at each perforation level, it concerns the accuracy and reproducibility of the results. Conversely, if the deviations between the methods remain relatively small and stable across all perforation percentages, this consistency suggests that the obtained values are robust and reliable for scientific interpretation. To quantitatively analyze these variations, the percentage difference between the methods was calculated for each perforation area.

Table 9. The percentage reductions in Young's modulus of the perforated 316L stainless steel plates obtained by the method of determination of resonant frequencies, the three-point flexural test method, and the finite element method (FEM).

Perforation Area (%)	Percentage Reduction in Young's Modulus Obtained by the Method of Determination of Resonant Frequencies (%)	Percentage Reduction in Young's Modulus Obtained by the Three-Point Flexural Test (%)	Percentage Reduction in Young's Modulus Obtained by FEM (%)
3.14	0	0	0
4.91	2.18	7.51	4.54
8.78	7.46	10.45	11.05
19.63	23.65	26.68	29.52

Figure 11 presents a comparative analysis of the percentage reduction in Young's modulus obtained by three methods across different perforation levels. The results show that, at 3.14% perforation, all methods provide nearly identical values, confirming consistency. As perforation increases, the values begin to diverge. At 4.91%, FEM predicts the highest reduction, while the resonant frequency and flexural test methods show slightly lower drops. At 8.78%, the difference becomes more distinct, with FEM showing an 11.05% reduction, resonant frequency showing 7.46%, and the flexural test showing 10.45%. At the highest perforation (19.63%), all methods show a significant drop, as follows: FEM (29.52%), resonant frequency (23.65%), and flexural test (26.68%). The convergence of the results at this stage suggests that all methods reflect the same overall mechanical trend, despite earlier discrepancies.

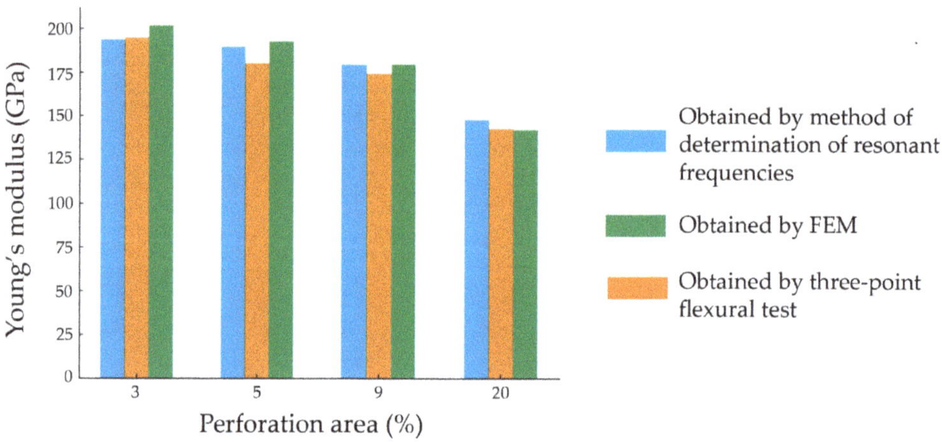

Figure 11. Dependence of the Young's modulus of the perforated 316L stainless steel plates on the perforation area, obtained by the method of determination of resonant frequencies, the three-point flexural test method, and the FEM.

At a perforation area of 20%, a more pronounced divergence appears, likely due to the increasing complexity of local deformation mechanisms and potential limitations in the FEM mesh resolution and boundary conditions at high porosity levels. Across all perforation levels, the three-point bending test results show slightly lower Young's modulus values, which can be attributed to the inherent differences between static flexural loading and dynamic resonant methods, as well as the influence of shear effects in flexural testing.

The observed non-linear decrease in Young's modulus with increasing perforation is primarily attributed to the complex interaction between hole diameter and spacing (pitch). While both parameters influence the overall reduction in stiffness, the effect is not strictly proportional to the perforated area alone. As perforation increases, stress concentrations and local deformation zones around the holes interact, especially when the spacing becomes comparable to the hole diameter, which leads to an accelerated reduction in effective stiffness. The results of the polynomial regression analysis further indicate that spacing (pitch) has a more significant influence on the modulus than hole diameter, particularly due to its role in controlling the degree of ligament connectivity between holes. Smaller pitch values lead to reduced load-bearing paths and a more pronounced decline in stiffness, explaining the non-linear trend observed in the experimental data.

5. Conclusions

This research displays a novel approach for investigating the elastic properties of perforated metal plates using a non-destructive method of determining resonance frequencies with a laser vibrometer.

This research discovered that an increase in perforation area leads to a reduction in the Young's modulus of 316L stainless steel plates.

For 316L stainless steel plates (perforated plates dimensions of 50 mm in height, 20 mm in width, and 1 mm in thickness; hole diameter of 1 mm; and pitch between the holes of 2, 3, 4, and 5 mm), increasing the perforation area from 3.14% to 19.63% led to a decrease in the Young's modulus of the perforated plates from 193.63 GPa to 147.8 GPa, which was assessed using a non-destructive approach based on resonant frequency measurements obtained via laser vibrometry.

Additional results were provided by a static three-point bending test, which showed a similar downward trend in Young's modulus with an increasing perforation area—from 194.4 GPa at 3.14% perforation to 142.6 GPa at 19.63%.

Furthermore, the finite element method (FEM) simulations confirmed that variations in perforation geometry—specifically, larger hole diameters and reduced pitch—result in a notable decline in the Young's modulus, with values ranging from 201.5 GPa down to 72.6 GPa, depending on the configuration (diameter of perforation and pitch between perforations).

The results of the polynomial regression analysis of the virtual experiment confirmed that the effective Young's modulus is affected by the perforation diameter and the pitch between perforations in a non-linear and interdependent way. This supports the conclusion that the perforation area alone is insufficient to fully characterize the elastic behavior of perforated plates.

A static analysis of the experimental results showed that the relationship between perforation geometry and elastic properties is non-linear and can be accurately approximated using a second-order polynomial model.

The reduction in Young's modulus of perforated 316L stainless steel plates also impacts their strength properties. It is, therefore, recommended that future research should explore the mechanical, strength, and fatigue properties of perforated 316L stainless steel plates in order to identify the potential risks associated with increased perforation area and to determine the optimal perforation areas for these materials for future use in orthopedic implants.

To achieve an effective Young's modulus below 70 GPa, a perforation ratio above 55%, with a perforation diameter exceeding 2.5 mm and pitch values below 3 mm, is recommended. The results demonstrate that, with the current material, further reduction in the effective Young's modulus is limited by technological constraints on maximum perforation diameter and minimum pitch. However, by selecting materials with intrinsically lower Young's modulus values (e.g., Ti-6Al-4V), it would be possible to achieve target values suitable for orthopedic applications. Moreover, further optimization of hole arrangement and perforation pattern could enable additional tailoring of elastic properties, although this requires a dedicated investigation and will be addressed in future studies.

To further validate the applicability of the developed approach for orthopedic use, future studies should include mechanical testing in simulated physiological environments, such as resonance frequency measurements in PBS solution at body temperature (37 °C), to assess the possible effects of fluid interaction on the elastic properties.

The obtained results clearly indicate a consistent reduction in the Young's modulus of 316L stainless steel perforated plates with an increasing perforation area and variation in perforation geometry. This suggests that the elastic properties of orthopedic implant structures can be deliberately tailored through controlled perforation, offering a viable strategy to affect the stress shielding effect by more closely aligning the implant's elastic properties with the elastic properties of the bone tissue.

Author Contributions: Conceptualization, M.O. and V.M.; methodology, M.O., V.M. and P.A.; software, P.A. and M.O.; investigation, M.O., V.M., V.Z. and P.A.; data curation, M.O., V.M., P.A. and D.S.; writing—original draft preparation, M.O., V.M. and D.S.; writing—review and editing, M.O., V.M.,

D.S. and P.A.; visualization, P.A. and M.O.; project administration, V.M., V.Z. and M.O. All authors have read and agreed to the published version of the manuscript.

Funding: This research was funded by the project of the Latvian Council of Science LZP FLPP no. LZP-2021/1-0290 "Comprehensive assessment of the condition of bone and muscle tissue using quantitative ultrasound (BoMUS)".

Data Availability Statement: Data supporting the results presented can be provided upon request to the respective author.

Conflicts of Interest: The authors declare no conflicts of interest.

References

1. Nuwirth, A.L.; Ashley, B.S.; Hardaker, W.M.; Sheth, N.P. *Biomedical Applications of Metals*; Rai, M., Ingle, A.P., Medici, S., Eds.; Springer International Publishing: Cham, Switzerland, 2018; pp. 3–95.
2. Swee, H.T. *Engineering Materials for Biomedical Application*; World Scientific Publishing Co. Pte. Ltd.: Hackensack, NJ, USA, 2004; pp. 4–29.
3. Marco Specialty Steel, Inc. Designers, Specifiers & Buyers. In *Handbook for Perforated Metals*; Industrial Perforators Association: Houston, TX, USA, 1993; pp. 3–14.
4. Mironovs, V.; Lisicins, M. *Perforated Metal Materials and Their Use Possibilities*; RTU Publishing House: Riga, Latvia, 2015; pp. 15–152.
5. Lisicins, M.; Serdjuks, D.; Akishin, P.; Mironovs, V.; Goremikins, V.; Lapkovskis, V. Improving the Structural Efficiency of Punched Metal-Material-Based Composites. *Polymers* **2024**, *16*, 3468. [CrossRef] [PubMed]
6. Baik, S.C.; Oh, K.H.; Lee, D.N. Analysis of the Deformation of a Perforated Sheet under Uniaxial Tension. *J. Mater. Process. Technol.* **1996**, *58*, 139–144. [CrossRef]
7. Xinjian, D.; Arnaud, W.; Wilkinson, D.S.; Metzger, D.R. Plastic Limit Analysis of Perforated Material Under Finite Deformation. In Proceedings of the ASME 2005 Pressure Vessels and Piping Conference, Denver, CO, USA, 17–21 July 2005; Volume 2, pp. 1–5.
8. Wadley, H.N. Cellular Metals Manufacturing. *Adv. Eng. Mater.* **2002**, *4*, 726–733. [CrossRef]
9. Qian, Y.J.; Kong, D.Y.; Liu, S.M.; Sun, S.M.; Zhao, Z. Investigation on micro-perforated panel absorber with ultra-micro perforations. *Appl. Acoust.* **2013**, *74*, 931–935. [CrossRef]
10. Abo Dorra, H.M.; Abd-Elbasseer, M.; Mahmoud, A.E.A.; Mohamed, H.K. Effect of Orifices Distribution Pattern on the Sound Absorption Performance of Double Metal Micro-Perforated Panels. *Int. J. Mech. Prod. Eng. Res. Dev.* **2020**, *10*, 15467–15476.
11. Mironovs, V.; Kuzmina, J.; Serdjuks, D.; Usherenko, Y.; Lisicins, M. Sustainable Lifecycle of Perforated Metal Materials. *Materials* **2023**, *16*, 3012. [CrossRef] [PubMed]
12. Karakaya, C. Numerical Investigation on Perforated Sheet Metals under Tension Loading. *Open Chem.* **2022**, *20*, 244–253. [CrossRef]
13. Reza, F.; Tamaki, Y.; Takahashi, H.; Iwasaki, N.; Miyazaki, T. Properties of a gypsum-bonded magnesia investment using a K_2SO_4 solution for titanium casting. *Dent. Mater. J.* **2009**, *28*, 4. [CrossRef]
14. Kroes, L. Creating More Efficiency and Patient Safety by Changing Processes and Contents of Instrument Trays. Master's Thesis, University of Twente, Enschede, The Netherlands, 2009.
15. Wexler Surgical. Available online: https://www.wexlersurgical.com/wexler-metal-sterilization-tray-single-level-wsilicone-brackets-105quotx215quotx25quot-266x533x63cm-p-1464.html (accessed on 1 October 2024).
16. Dewidar, M.; Khalil, A.; Lim, J.K. Processing and Mechanical Properties of Porous 316L Stainless Steel for Biomedical Applications. *Trans. Nonferrous Met. Soc.* **2007**, *17*, 468–473. [CrossRef]
17. Shahzamanian, M.M.; Banerjee, R.; Dahotre, N.B.; Srinivasa, A.R.; Reddy, J.N. Analysis of stress shielding reduction in bone fracture fixation implant using functionally graded materials. *Compos. Struct.* **2023**, *321*, 117262. [CrossRef]
18. Wintermantel, E.; Mayer, J.; Goehring, T.N.; Aqida, S. Composites for Biomedical Applications. *Ref. Modul. Mater. Sci. Mater. Eng.* **2016**, 2–3. [CrossRef]
19. William, D.F. *Handbook of Biomaterial Properties*, 2nd ed.; Murphy, W., Black, J., Hastings, G., Eds.; Springer: New York, NY, USA, 2016; pp. 153–154.
20. Ratner, B.D.; Hoffman, A.S.; Schoen, F.J.; Lemons, J.E. *Biomaterials Science. An Introduction to Materials in Medicine*, 2nd ed.; Elsevier: London, UK, 2004; pp. 526–528.
21. Follansbee, P.S. *Fundamentals of Strength. Principles, Experiment, and Applications of an Internal State Variable Constituve Formulation*; John Wiley & Sons: Hoboken, NJ, USA, 2014; pp. 7–8.
22. Callister, W.D. *Fundamentals of Materials Science and Engineering: An Integrated Approach*, 2nd ed.; John Wiley & Sons: Hoboken, NJ, USA, 2005; pp. 183–190.

23. Callister, W.D.; Rethwisch, D.G. *Fundamentals of Materials Science and Engineering: An Integrated Approach*, 3rd ed.; John Wiley & Sons: Hoboken, NJ, USA, 2008; pp. 226–228.
24. Ahmed, M.S. Numerical Analysis of the Perforated Steel Sheets Under Uni-Axial Tensile Force. *Metals* **2019**, *9*, 632. [CrossRef]
25. Callister, W.D.; Rethwisch, D.G. *Materials Science and Engineering: An Introduction*, 10th ed.; John Wiley & Sons: Hoboken, NJ, USA, 2018; pp. 433–435.
26. Migonney, V. *Biomaterials. Bioengineering and Health Science Series*; ISTE Ltd.: Hertfordshire, UK; John Wiley & Sons: Hoboken, NJ, USA, 2014; pp. 31–34.
27. Ratner, B.D.; Hoffman, A.S.; Schoen, F.J.; Lemons, J.E. *Biomaterials Science. An Introduction to Materials in Medicine*, 1st ed.; Academic Press: Millbrae, CA, USA, 1996; pp. 37–50.
28. Hanawa, T. Research and development of metals for medical devices based on clinical needs. *Sci. Technol. Adv. Mater.* **2012**, *13*, 064102. [CrossRef]
29. Sáenz de Viteri, V.; Fuentes, E. Titanium and Titanium Alloys as Biomaterials. In *Tribology—Fundamentals and Advancements*; Gegner, J., Ed.; IntechOpen: Rijeka, Croatia, 2013; pp. 160–163.
30. Krasnikovs, A.; Kononova, O.; Machanovskis, A.; Zaharevskis, V.; Akishins, P.; Ruchevskis, S. Characterization of mechanical properties by inverse technique for composite reinforced by knitted fabric. Part 2. Experimental evaluation of mechanical properties by frequency eigenvalues method. *J. Vibroeng.* **2012**, *14*, 691–698.
31. Kovalovs, A.; Rucevskis, S. Identification of elastic properties of composite plate. *IOP Conf. Ser. Mater. Sci. Eng.* **2011**, *23*, 012034. [CrossRef]
32. Wesolowski, M.; Barkanov, E.; Rucevskis, S.; Chate, A.; La Delfa, G. Characterisation of elastic properties of laminated composites by non-destructive techniques. In Proceedings of the ICCM International Conferences on Composite Materials, Edinburgh, UK, 27–31 July 2009; pp. 5–7.
33. Rothberg, S.J.; Allen, M.S.; Castellini, P.; Di Maio, D.; Dirckx, J.J.J.; Ewins, D.J.; Halkon, B.J.; Muyshondt, P.; Paone, N.; Ryan, T.; et al. An international review of laser Doppler vibrometry: Making light work of vibration measurement. *Opt. Lasers Eng.* **2017**, *99*, 11–22. [CrossRef]
34. Barkanov, E.; Wesolowski, M.; Akishin, P.; Mihovski, M. Techniques for Non-Destructive Material Properties Characterisation. In *Non-Destructive Testing and Repair of Pipelines*; Barkanov, E.N., Dumitrescu, A., Parinov, I.A., Eds.; Engineering Materials; Springer: Cham, Switzerland, 2018; pp. 194–203.
35. Rikards, R.; Abramovich, H.; Green, T.; Auzins, J.; Chate, A. Determination of Elastic Properties of Stiffened Composite Shells by Vibration Analysis. *Mech. Adv. Mater. Struct.* **2003**, *10*, 177–180.
36. Pickett, G. Equations for Computing Elastic Constants from Flexural and Torsional Resonant Frequencies of Vibration of Prisms and Cylinders. *Proc. ASTM* **1945**, *45*, 846–865.
37. *ASTM Standard E 1876-07*; Standard Test Method for Dynamic Young's Modulus, Shear Modulus and Poisson's Ratio by Impulse Excitation of Vibration. ASTM International: West Conshohocken, PA, USA, 2007.
38. Losertová, M.; Štamborská, M.; Lapin, J.; Mareš, V. Comparison of deformation behavior of 316L stainless steel and Ti6Al4V alloy applied in traumatology. *Metalurgija* **2016**, *55*, 667–670.
39. Auzins, J.; Janushevskis, A.; Janushevskis, J.; Skukis, E. Software EDAOpt for experimental design, analysis and multiobjective robust optimization. In Proceedings of the 1st International Conference on Engineering and Applied Sciences Optimization, Kos, Greece, 4–6 June 2014; pp. 1055–1077.
40. Yaqoob, K.; Amjad, I.; Awan, M.A.M.; Liaqat, U.; Zahoor, M.; Kashif, M. Novel Method for the Production of Titanium Foams to Reduce Stress Shielding in Implants. *ACS Omega* **2023**, *8*, 1876–1884. [CrossRef]
41. Niinomi, M.; Nakai, M. Titanium-Based Biomaterials for Preventing Stress Shielding Between Implant Devices and Bone. *Int. J. Biomater.* **2011**, *2011*, 836587. [CrossRef] [PubMed]

metals MDPI

Article

Acoustic Emission During Rubber-like Deformation in $Ni_{51}Fe_{18}Ga_{27}Co_4$ Single Crystalline Shape Memory Alloys

Lajos Daróczi [1], Sarah M. Kamel [1,2], László Z. Tóth [1], Elena Yu. Panchenko [3], Yuri I. Chumljakov [3] and Dezső L. Beke [1,*]

[1] Department of Solid State Physics, Doctoral School of Physics, Faculty of Science and Technology, University of Debrecen, P.O. Box 400, 4002 Debrecen, Hungary; lajos.daroczi@science.unideb.hu (L.D.); sarahkamel@science.unideb.hu (S.M.K.); toth.laszlo@science.unideb.hu (L.Z.T.)
[2] Physics Department, Faculty of Science, Ain Shams University, Abbassia, Cairo 11566, Egypt
[3] Siberian Physical Technical Institute, Tomsk State University, Tomsk 634050, Russia; panchenko@mail.tsu.ru (E.Y.P.); chum@phys.tsu.ru (Y.I.C.)
* Correspondence: dbeke@science.unideb.hu

Abstract: Acoustic emission, AE, belonging to a rubber-like deformation in a martensitic state after the stabilization aging of the stress-induced martensite (SIM aging) of $Ni_{51}Fe_{18}Ga_{27}Co_4$ single crystals in compression, were investigated. AE activity in the plateau regions of the stress–strain loop is due to a massive reorientation from the variants produced by SIM aging to the variants preferred by the compressive stress (perpendicular to the stress used in SIM aging) and vice versa. For unloading, the large AE activity just at the knee point of the stress–stain curve is attributed to the difficulty of the re-nucleation of the SIM aging-stabilized martensite variant. The amplitude, peak energy, and area of signals can be described by power-like distributions and the characteristic exponents are in good agreement with data obtained in other alloys. Power law cross-correlations between the energy, E, and amplitude, A, as well as between the area, S, and the amplitude, A, were also analyzed. It was found that the exponents are given by $3 - \varphi$ as well as $2 - \varphi$, respectively, with $\varphi \cong 0.7$. Normalized universal temporal shapes of avalanches (i.e., the $\frac{U}{A}$ versus $\frac{t}{A^{1-\varphi}}$ plots, where U is the detected voltage) for a fixed area scale very well together. The tail of the normalized temporal shape decays more slowly than the theoretical prediction, which can be attributed to an intrinsic absorption of AE signals and/or to the overlap of sub-avalanches.

Keywords: rubberlike effect; martensite stabilization; acoustic emission; noise analysis

1. Introduction

The deformation of shape memory materials in a martensitic state can result in a large residual strain due to the rearrangement of martensite variants (a superplastic behavior) into a single variant structure preferred by the applied stress [1]. Aging in a martensitic state under uniaxial stress (also called the stabilization aging of stress-induced martensite or SIM aging) [2–10] leads to a stabilization of the induced martensite variant and its reorientation (under a stress perpendicular to the direction applied during SIM aging) and shows a hysteresis on the stress–strain curve without a significant residual strain (rubber-like behavior) [2]. Thus, this rubber-like behavior, compatible with the two-way shape memory recoverable strain of the usual shape memory alloys can have important practical applications due to the large reversible shape change induced either by the uniaxial stress or magnetic field in ferromagnetic shape memory alloys [6–10]. The most

https://doi.org/10.3390/met15020131

important mechanism of the variant rearrangement is the twin boundary motion, which has a discontinuous character, leading to acoustic emission, AE, signals [11–14].

The above intermittent character is manifested in the emission of signal avalanches, which have characteristic statistical behavior. The probability distribution density function of a typical signal parameter, x (like the peak amplitude, A, area, S, or energy, E), according to the tuned criticality behavior [15], follows a truncated power law form [12,15]:

$$P(x) \propto x^{-\mu} \exp\left(-\frac{x}{x_c}\right) \tag{1}$$

where μ is the characteristic exponent and x_c is the cutoff value. Here, x denotes a measured parameter of the avalanche, like the amplitude, A, area, S, or energy, E, or duration, D. Furthermore, well-known power law relations exist between the measured avalanche parameters and, e.g., the mean field theory, MFT, [15–17] predicts that $S \sim D^{\gamma}$ with $\gamma = 2$ and $S \propto A^2$, $E \propto A^3$. Such power relations are consequences of the self-similarity of the crackling noises [15].

It is well known that due to experimental difficulties (finite threshold values [18, 19] and transfer distortions of the original source signal [18–20]), most of the measured parameters cannot reflect the features of the source (i.e., the rate of the intermittent local shape changes during twin reorientations). For example, the duration time, D, (and related parameters containing D in their definitions) for AE avalanches can be strongly polluted by the internal absorption of the AE waves (which can be characterized by an exponential expression, $e^{-\frac{t}{\tau_a}}$, where τ_a is the attenuation time) [18,19]. It was shown in [19] that, although the exponents of the probability distribution functions can be more or less free of such effects, the scaling relations between the measured parameters can be strongly influenced. Moreover, it was also pointed out in [19] that the correlations between S and A as well as E and A can be characterized by exponents considerably less than the values given by the MFT. This is called "enigma" for acoustic emission [19]. In a recent communication [18], the above enigma was investigated by assuming a power relation between A and the rising time, R, of the AE signal ($R \propto A^{1-\varphi}$), and it was found that $S \propto A^{2-\varphi}$ and $E \propto A^{3-\varphi}$ holds, respectively. The values of φ were the same for the intermittent motions of interfaces between martensite and austenite in two different shape memory single crystals, (of $Ni_{45}Co_5Mn_{36.6}In_{13.4}$ and $Ni_{49}Fe_{18}Ga_{27}Co_6$ alloys) with $\varphi = 0.8 \pm 0.2$ for both alloys.

It is also well known that the above-mentioned exponents in Equation (1) are quite robust, and change only slightly for different mechanisms [18,21]. Thus, the focus is increasingly on the average temporal shapes of avalanches, i.e., the average of the detected voltage signal ($U(t)$) of the magnetic or acoustic emission for the fixed size or duration [16,17,20,22], which can be better related to the given type of the mechanism [16,17]. Regarding the average avalanche shape for fixed area, S, it was obtained from the theory [22] that

$$U(t) \sim < v(T) >_S = ate^{-\left(\frac{t}{\tau}\right)^2} \tag{2}$$

where $v(T)$ and $U(t)$ are the interface velocity and the detected voltage. a and τ are non-universal (material-dependent) constants: τ is the characteristic time of the avalanche decay [23]. On this basis, the self-similarity of the emitted avalanches [21] is expected, and using appropriate scaling parameters, the normalized $U(t)$ functions should fall on a common curve for different bins of S. For the experimental verification, whether an appropriate scaling can indeed lead to good scaling together, dividing both axes by $S^{0.5}$ (based on the power relations predicted by MFT, i.e., $A \propto S^{0.5}$ as well as $D \propto A$) was used [20,23,24]. However, no good scaling was achieved, and the curves did not scale completely in a universal way. As it was mentioned above, there exists a power relation between A and R as well, as follows: $R \propto A^{1-\varphi}$ [18]. Thus, by scaling the voltage by A and the time by $A^{1-\varphi}$, nice common functions were obtained not only for the motion of interfaces between austenite and martensite in $Ni_{49}Fe_{18}Ga_{27}Co_6$ single crystals [18], but for the motion of the individual twin boundary (of type I) in Ni-Mn-Ga single crystals [25] too.

This paper, we present experimental results on AE measurements during rubber-like deformation (i.e., during twin rearrangements) of $Ni_{51}Fe_{18}Ga_{27}Co_4$ ferromagnetic shape memory single crystals. According to the best of our knowledge, this is the first complete AE investigation for the whole cycle of such a large (about 15%) rubber-like recoverable strain. In addition to the determination of the characteristic exponents in Equation (1), the scaling relations between the area and peak amplitude as well as between the energy and peak amplitude are determined and the obtained power exponents of these are used to find normalization factors for the construction of universal average temporal avalanche shapes. Our results will also be compared with AE results obtained during the stress- or magnetic field-induced motion of single twin boundaries or different ensembles of twin boundaries in Ni_2MnGa alloys. Universal averaged temporal avalanches are also constructed at a fixed area, by dividing the voltage and time scales by A as well as by $A^{1-\varphi}$, using the φ value calculated from the exponents of power relations between the S and A as well as between the E and A.

2. Experimental

The $Ni_{51}Fe_{18}Ga_{27}Co_4$ single crystals were prepared using vacuum induction melting and the single crystals were grown by the Bridgman method in a helium atmosphere at Tomsk University, Russia. Dimensions of the sample were 3 × 3 × 5 mm. Homogenization heat treatment was performed at 1175 °C for 1 h, which was followed by a quench to room temperature. The B2 high-temperature phase of the quenched crystals undergoes a martensitic transformation to the L10-martensite. SIM (stress-induced martensite) stabilization aging was performed under compression along the $[110]_A$ direction in the martensitic state at 150 °C for 1 h.

Compression tests were made by an Instron 4465 tensile test machine at room temperature along the $[001]_A$ direction. Acoustic emission (AE) measurements were performed by a piezoelectric sensor (MICRO-100S from Physical Acoustic Corporation, Princeton Junction, NJ, USA), attached to the steel compression anvil (Figure 1). The microphone had a good frequency response between 100 KHz and 1 MHz, which is nearly flat in the range between 0.2 and 1 MHz (at about −75 dB in average) with maximum ± 10 dB deviation (1 V reference value). The microphone was integrated with a 30 dB pre-amplifier. Furthermore, we connected a primary amplifier to the microphone with logarithmic gain, offering a dynamic range of 90 dB and operating within a band pass frequency range of 30 KHz to 1 MHz. For the conditioning and acquisition of acoustic data, a Sensophone (Geréb and Co GmbH, Budapest, Hungary) AE measurement system was used with a 8 MS/s sampling rate (for the details, see [26]). In the evaluation of acoustic signals, a 0.04 V threshold value and a 4×10^{-5} s hit detection time was used.

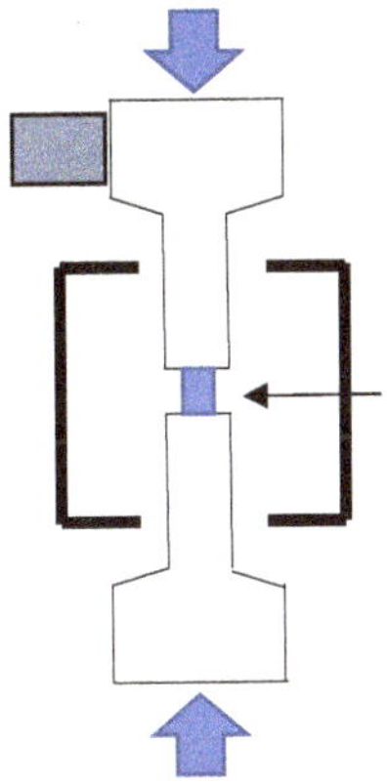

Figure 1. Schematic picture of the measuring setup. The arrow shows the sample and the box at the top anvil is the sensor.

3. Results

The stress–strain curve, taken with a strain rate of 1.25 × 10−3 s−1 along the $[001]_A$ direction (which is perpendicular to the $[110]_A$ direction of SIM aging) during the rubber-like behavior of the $Ni_{51}Fe_{18}Ga_{27}Co_4$ sample, is presented in Figure 2a. The cycles were stable and curves of more than 10 cycles were fully collapsed. As the result of SIM aging (see, e.g., [7]), the recoverable strain is around 15%. Figure 2b shows the stress curves, and the simultaneously measured AE signals as well as the cumulative energy of the AE signals ($\sum E = E_i$) as the function of time. The number of AE events remarkably increases on the unloading curve (at around 125 s on Figure 2b), with a large jump in $\sum E$.

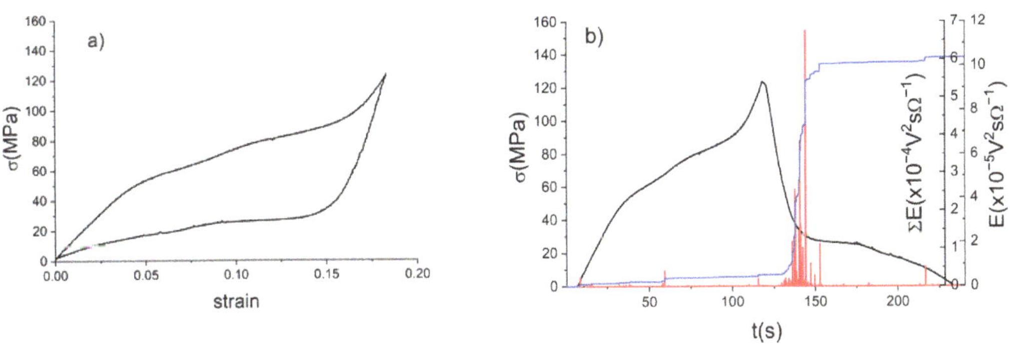

Figure 2. (**a**) Compression stress–strain curve; (**b**) stress (black), energy of acoustic emission signals (red), and the cumulative AE energy, $\sum E$, (blue) as a function of time (taken with a 1.25 × 10^{-3} s^{-1} strain rate). A large acoustic activity and large jump in $\sum E$ can be observed on the unloading curve.

The AE data were used for the calculation of the characteristic exponents in Equation (1) for the area, S, energy, E, and amplitude, A. The S, E, and A of acoustic avalanches were calculated from the detected voltage, $U(t)$ (temporal avalanche shape) according to their usual definitions [25]: A is the maximum value, $S = \int_0^D U(t)dt$, and $E = \frac{1}{R_o}\int_0^D U^2(t)dt$, where R_o is an arbitrarily chosen resistance (R_o = 1 MΩ).

In order to get the exponents from Equation (1), the probability density functions, PDF, (calculated using a logarithmic boxing of the quantities $P(x_i) \sim \frac{N_i}{N}$, where N_i and N denote the number of events at a certain value of x_i and the number of all hits, respectively) were fitted using a two-parametrical nonlinear fitting by the Levenberg–Marquardt least squares method. As an illustration, Figure 3 shows the energy probability distribution density of the acoustic signals, averaged for both directions. It can be seen that a straight line can very well fit the points over about four–five orders of magnitude. The maximum likelihood fit is a recommended method to check the reliability of exponents calculated from the log P versus log E plots [27]. We obtained that, in all cases, the exponents determined by the two methods were the same within the error bars given in Table 1 (see also Figure 3a,b). Table 1 shows the calculated exponents for energy, amplitude, and area. We have carried out a sensitivity analysis (always considering the corresponding maximum likelihood fits as well), regarding the small upward curvature in Figure 3a. This is sensitive to the choice of the threshold. The plot shown is the optimum fit with a 0.04 V threshold value.

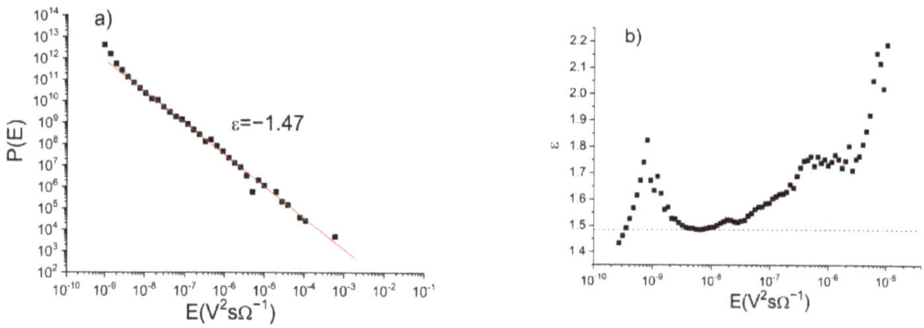

Figure 3. The energy probability distribution density function of acoustic signals obtained during rubber-like deformation (**a**). (**b**) The maximum likelihood fit for the exponent.

Table 1. Power exponents of the acoustic emission parameters.

P(E)~$E^{-\varepsilon}$	P(A)~$A^{-\alpha}$	P(S)~$S^{-\delta}$	E~$A^{3-\varphi}$	S~$A^{2-\varphi}$
ε	α	δ	φ	φ
1.5 ± 0.1	2.3 ± 0.2	1.6 ± 0.1	0.7 ± 0.1	0.7 ± 0.1

Figures 4 and 5 show the power relations between the energy and amplitude as well as the area and amplitude, respectively. It can be seen that, in addition to the first parts, belonging to small amplitudes, where the amplitude is comparable to the threshold (0.04 V) and a downward curvature is observed (see also Figure 6 and Figure 7 in [18] or Figure 2d in [25]), a straight line can be fitted. As it was shown in [26], such curvature appears if the amplitude per threshold ratio is less than 10. Using the prediction of [18], the exponentsfor the power relations between E and A as well as S and A can be expressed as $\mathbf{3}-\boldsymbol{\varphi}$ as well as $\mathbf{2}-\boldsymbol{\varphi}$, respectively. Table 2 also contains the values of φ obtained from our measurements.

It is worth mentioning that the data collected during the forward and reverse deformation were collapsed into one file; thus, the above exponents were calculated using all (up and down) of the points. This is in contrast to the results obtained from austenite/martensite transformations in different shape memory alloys, where frequently the exponents for the up and down processes were different [28]. We will turn to this below, commenting on the difference of the total numbers ($\sum N = \sum_i N_i$) and total AE energy ($\sum E = \sum_i E_i$) during loading and unloading branches and we will discuss the fine details of the obtained AE data.

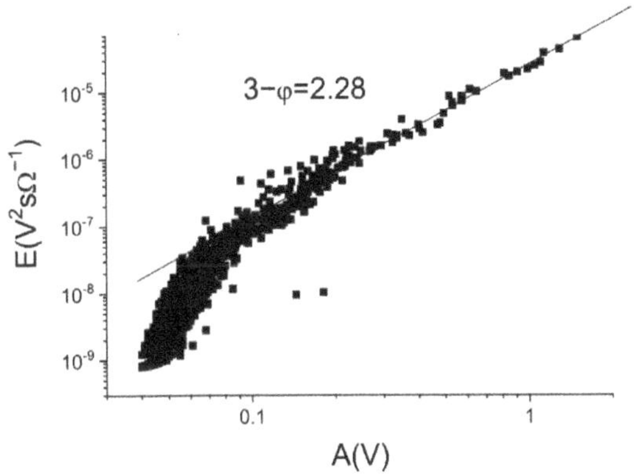

Figure 4. *LogE* versus *LogA*. The insert shows the slope of the straight line fitted.

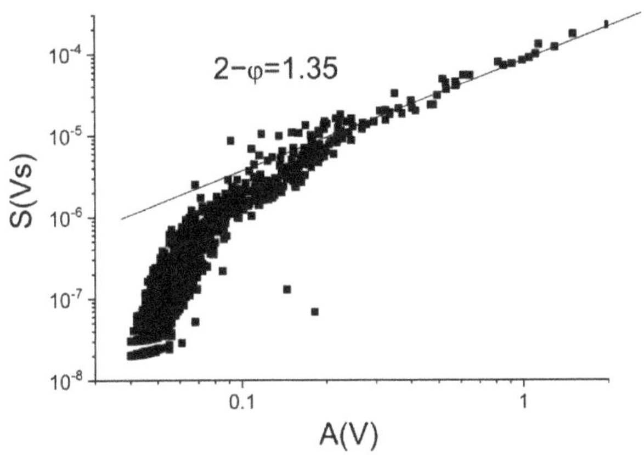

Figure 5. *LogS* versus *LogA*. The insert shows the slope of the straight line fitted.

4. Discussion

4.1. AE During Loading and Unloading

The origin of the rubber-like behavior is the so-called symmetry conforming short-range ordering [29] during SIM aging (which leads to a diffusional stabilization of the given single variant martensite, preferred by applied stress). It is not surprising that the recoverable strain is large for our $Ni_{51}Fe_{18}Ga_{27}Co_4$ single crystals since, in this case, the stabilized martensite structure is dominantly a single martensite variant, and the aging temperature and time of SIM aging was high enough to reach the complete short-range order according to this martensite structure [7]. The single variant martensite reorients during the application of the compressive uniaxial stress (in $[001]_A$ direction, perpendicular to the $[110]_A$ SIM aging direction of austenite) and if it is a complete reorientation then a recoverable strain close to the theoretical value can be reached, which is about 16.6% in this alloy [7]. In our case, the experimental value is about 15% (see Figure 2a), which indicates that the full potential of the transformation strain was not realized in our experiment. This can be related to the fact that the SIM-aged state or the reoriented state (or both) is not a fully de-twinned structure of the corresponding martensite (see also Figure 5). Here, we use the notations of [7] and [30] for this alloy: the dominant variant V1 ($[001]_A || [001]_M$) is formed during SIM aging, while V2 ($[010]_A || [001]_M$) and V3 ($[100]_A || [001]_M$) are the reoriented variants [30] (preferred by the stress under which the rubber-like behavior is investigated. see also Figure 3a in [7], as well as Figure 2 in [30]) The presence of variants V2 and V3 is the indication of the incomplete de-twinning [30]. Nevertheless, during loading reorientation from the (initially dominant) V1 to V2/V3, it takes place and the reveres process sets in during unloading.

It can be seen from Figure 2 that, for loading, there are some AE signals already before the start of the plateau (at about 3.5% strain), but these most probably can be attributed to micro-friction in the course of matching the sample to the testing machine [31]. The sporadic AE activity and a slight gradual increase of the cumulative AE energy in the whole remaining part of the loading stress–strain/time curve belongs to the reorientation process from V1 to V2/V3. For the unloading part, the situation is different in the sense that here, a large number of AE signals and a large jump in $\sum E$ appear at the beginning of the plateau (at around 125 s), which is followed by a similar lower activity (like along the loading branch) until the end. In this latter plateau region, a jerky rearrangement of the variant structure takes place, either by the nucleation of the V2/V3 variant or by the motion of the already existing boundaries between the V1 and V2/V3 boundaries. The above picture is also in a good qualitative agreement with the simulation results on dynamical tweed, mobile kinks, and movement in the twin walls in ferroelastic materials [32,33]. The amplitude of jumps was higher in the yield regime during nucleation (at around the beginning of the plateau in our case) as compared to the plateau region where the transition between V1 and V2/V3 took place.

The high local activity on the unloading curve also leads to a remarkable difference in the total number of acoustic events and in the total acoustic energy ($\sum_i N_i = 1250$ and 2500 as well as $\sum_i E_i = 0.5 \times 10^{-4}$ V^{2}s and 5.5×10^{-4} V^{2}s for loading and unloading, respectively), which suggests that there is an AE asymmetry. Since the nucleation processes typically provide high AE activity [23,32,33], we interpret the asymmetry and large local AE activity on the unloading branch by the difficulty of the re-nucleation of the initial, differently oriented martensite variant V1. The difficulty is related to the almost completely de-twinned structure of the initial state: from the twinned state (Figure 6b) (in which twinning is the manifestation of some local stress relaxations by the formation of twins), the formation of the stabilized de-twinned structure needs to overcome an energy barrier of an elastic origin. Furthermore, the remaining part reflects similarly for the loading branch, with the jerky propagation of the boundaries between the two variants. The relatively large dissipation (wide hysteresis loop in Figure 2a) can also be a fingerprint of the difficult re-nucleation of the V1 variant.

Figure 6. An optical micrograph of the martensite structure: (**a**) original, V1, state with a dominantly de-twinned region, with minor residual twins; (**b**) twinned, reoriented, V2/V3, martensite; and (**c**) the final state, which is equivalent to the initial state.

Thus, it seems that while the processes in the plateau are similar in both branches, the above asymmetry is an indication that there are no nucleation difficulties for the appearance/further growth of the V2/V3 variant in loading, but in unloading, the re-nucleation of V1 is difficult. Understanding the structural details of this calls for a detailed microscopic investigation. Nevertheless, the fact that the observed transformation strain is less than the theoretically predicted value also reflects that the V1 and V2/V3 end states should not be fully de-twinned variants.

The above asymmetry in $\sum N$ and $\sum E$ suggests that it is worth checking whether the exponents of signals collected from the large local activity region only for unloading are the same as the average values given in Table 1 or not. We will turn back to this point below.

4.2. Power Exponents of the Acoustic Emission Parameters

Firstly, we compare the exponents of the amplitude or energy probability density functions, with exponents given in previous publications obtained either from the magnetic field- or stress field-induced martensite variant rearrangements in ferromagnetic shape memory alloys [11,13,25,34–36]. Table 2 contains the summary of the measured exponents, with our value calculated from all of the measured points.

Table 2. Characteristic exponents of the energy (ε) and amplitude (α) distributions measured during the external field-induced rearrangements of martensite variants and/or during the jerky shift of individual twin boundaries.

Single Crystal	Summary	ε	α
$Ni_{51}Fe_{18}Ga_{27}Co_4$, this work, averaged	Rubber-like, Stress-induced	1.5 ± 0.1	2.3 ± 0.2
Ni_2MnGa [13]	Single twin boundary Type I, stress-induced		2.3 ± 0.2
Ni_2MnGa [13]	Single twin boundary Type II, stress-induced (less than one decade energy range)		3.0 ± 0.2
Ni_2MnGa [34]	Type II	1.5 ± 0.1	
Ni_2MnGa [35]	Type I	1.5 ± 0.1	
Ni_2MnGa [25]	Single twin boundary Type I, stress-induced	1.5 ± 0.1	2.1 ± 0.1
Ni_2MnGa [36]	After compression along the [100] direction, magnetic field-induced	1.5 ± 0.1	1.8 ± 0.2

In [13,25,35], the AE results for a type I twin boundary were analyzed in Ni_2MnGa, while in [36], rearrangements of a magnetic field-induced twin boundary network was investigated. In Ni_2MnGa single crystals, the same energy and amplitude exponents can be concluded (if we do not take into account the results of [13] for type II twin boundaries, where less than one order of a magnitude-wide energy range was covered), which can be summarized by $\varepsilon = 1.5 \pm 0.2$ and $\alpha = 2.1 \pm 0.2$, respectively. It can also be seen from Table 2 that our present results are in good agreement with the above average data.

Thus, we can also conclude that the energy and amplitude exponents were the same both for the stress- or magnetic field-induced motion of single boundary or for the motion of different ensembles of twin boundaries in two different ferromagnetic shape memory materials. This result is similar to the conclusion drawn on AE exponents (with slightly different average values as obtained for twin boundary motions) collected during the austenite/martensite transformations in Ni_2MnGa, $Ni_{51}Fe_{18}Ga_{27}Co_4$, and $Ni_{45}Co_5Mn_{36.6}In_{13.4}$ single crystals [26].

The above results nicely demonstrate the robustness of the characteristic exponents present in relations given by Equation (1). It can also be seen that the φ values, as expected [18,25], are the same as calculated either from the cross-correlation between E and A or S and A.

As it was mentioned already in the introduction, it is worth going beyond the comparison of the energy and amplitude exponents only to investigate the scaling properties of the temporal avalanche shapes at fixed areas. Before this, it is worth checking the exponents of the AE parameters collected after the separation of the AE signals belonging to the large local AE activity during the re-nucleation of the SIM ageing-stabilized variant, V1. We carefully carried out this analysis and found that all the energies given in Table 1 were the same for both the separated and the remaining signals within the error bars. In addition, in Figures 4 and 5, and on the corresponding plots after separation, not any branching was observed. As it has been shown in a series of papers [37–40], the multibranching (i.e., $E = s_i A_i^2$ and for i = 1 and 2, $s_1 \neq s_2$) stems from the overlapping of sub-avalanches within one acoustic emission signal and the extent of this overlap can be different for different avalanche mechanisms. It was also shown that this effect can be very useful in making distinctions between different avalanche mechanisms [37]. Thus, we can conclude that on the basis of our results, no distinction can be made between AE signals belonging to the re-nucleation of the V1 variant and to the massive reorientation of the two variants (in the plateau regions).

4.3. Normalized Temporal Avalanche Shapes for Fixed Duration and Area

The next step is the construction of the normalized (universal) temporal avalanche shapes for a fixed area. The normalized functions for fixed area, in accordance with [18] and [25], can be obtained from (2) by scaling the voltage scale with A and the time scale with $A^{1-\varphi}$, since according to [26] $R \propto A^{\varphi}$ with $\varphi = 0.7$ (see Table 1). In constructing the average avalanche shapes, firstly different avalanche size ranges (bins) were selected. Within each size range, the avalanches were averaged, and the common avalanche parameters (amplitude, size, energy, and rising time) were determined for the average function belonging to this bin (see Figure 7). As a second step these functions were scaled together by dividing the U axis by A and the t axis by $A^{1-\varphi}$, as it is shown in Figure 8. It can be seen that, as expected, the averaged curves are collapsed, showing the self-similarity of the avalanches over a wide size range. It is worth commenting on two characteristics of Figure 8: (a) fluctuations (i.e., only a smoothed curve would fit to the shape predicted from Equation (2)), and (b) the last part (above about a 300 reduced time) decays much more slowly than predicted by Equation (2). The fluctuating behavior can be a consequence of our relatively bad statistics as compared, e.g., with similar AE measurements on motion of individual twin boundary in Ni_2MnGa single crystals (the number of events was about five times larger in [25] than in here). Another possibility is the effect of mechanical oscillations in the otherwise monotonic driving force [41]; if the typical durations of avalanches is comparable to the period of oscillations of the testing machine, some oscillations can appear on the temporal avalanche shape (this effect is expected to be more expressed in experiments made on micro-pillars [41]). However, as compared to our measurements, in $Ni_{49}Fe_{18}Ga_{27}Co_6$ [18], the duration times of avalanches emitted during martensitic transformation covered a similar range and the testing machine was the same: no oscillations were observed on the averaged temporal shapes. Regarding the behavior b), it can be noted that while in [25], the tail of the normalized temporal shapes at fixed area followed Equation (2) ($\sim e^{-\frac{t^{*2}}{2}}$, $t^* = \frac{t}{A^{1-\varphi}}$) with exponent 2, as expected, for Ni_2MnGa [25] or in $Ni_{49}Fe_{18}Ga_{27}Co_6$ [18,25], here, an exponent of about 1 can be obtained (see Figure 9). This latter value is similar to the exponent obtained in [23] where it was mentioned that it can be related to the absorption of the acoustic signals [18,23] characterized by an $\sim e^{-\frac{t}{\tau_a}}$ type function. From such a fit, a $\tau_a \cong 150$ μs attenuation time can be estimated which is in reasonable agreement with data obtained from AE experiments in $BaTiO_3$ ($\tau_a = 20$ μs) [19]. However, in [23], it was also discussed as a still-open question, that the presence of a slower decay in the tail region can also be related to the contribution of sub-avalanches. This can lead to the Omori-type decay (postseismic slip), observed in earthquakes [42]): $\sim t^{-p}$ with $p \cong 1$. Figure 10 shows the $logU^*$ versus $logt^*$ plot and it can be seen that it can also be fitted with a straight line and the slope of this is $p = 0.91$. Deciding between the above two possible explanations calls for further investigations.

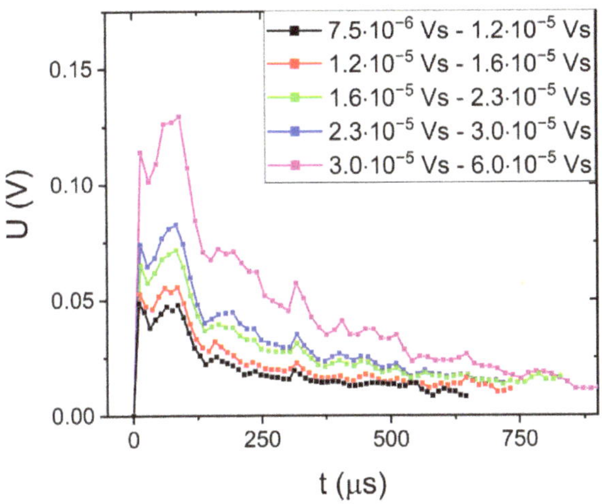

Figure 7. Averaged AE signals in different size (area) ranges (the insert gives the center values of S within which the averaging was made).

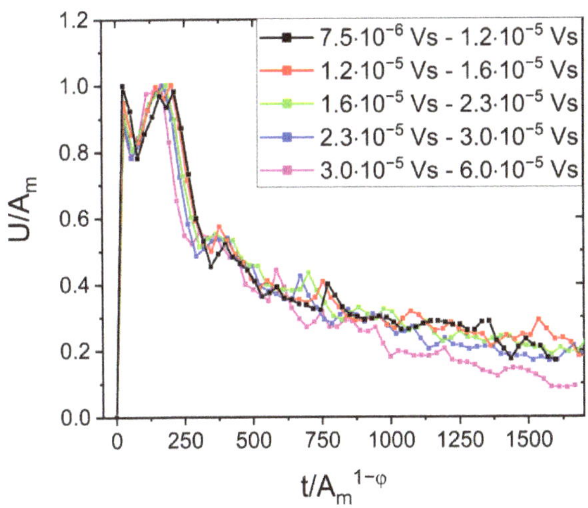

Figure 8. Averaged AE signals after a normalization of the vertical and horizontal axes by A_m and $A_m{}^{1-\varphi}$, respectively.

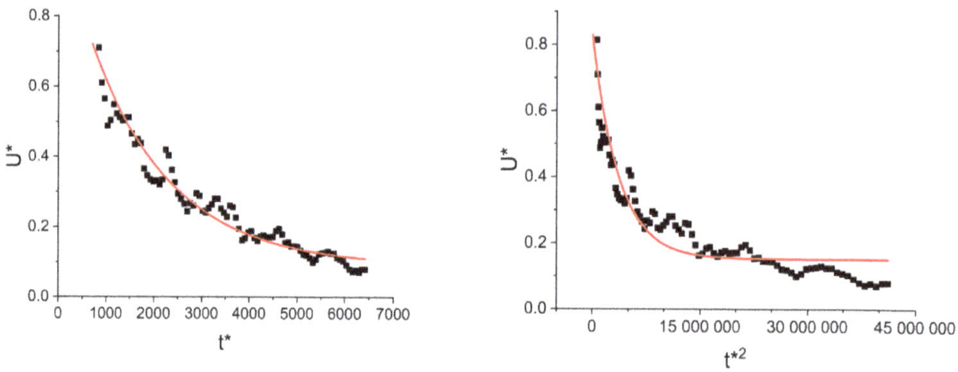

(**a**) (**b**)

Figure 9. Fitting the exponentially decaying tail of the curve on Figure 10 ($U^* = \frac{U}{A}$ and $t^* = \frac{t}{A^{1-\varphi}}$): according to Equation (2) (**a**) as well as by $\sim e^{-\frac{t}{\tau_a}}$ type function (**b**). It can be seen that the fit $\sim e^{-\frac{t}{\tau_a}}$ is better, and it yields $\tau_a \cong 150$ μs.

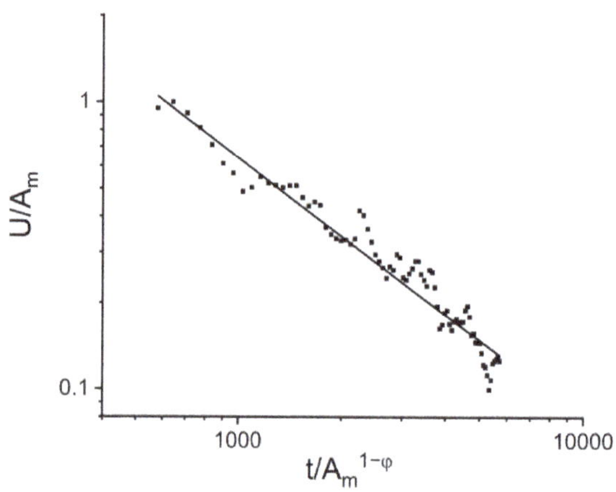

Figure 10. A $logU^*$ versus $logt^*$ plot from the tail region of Figure 8.

5. Conclusions

- It is shown that, after SIM aging, there is a well-detectable acoustic emission activity during the rubber-like deformation (up to about 15%) of $Ni_{51}Fe_{18}Ga_{27}Co_4$ single crystals.
- The observed large AE activity at the beginning of the unloading plateau on the stress–stain curve and the observed large AE asymmetry for loading and unloading is attributed to difficulties in the re-nucleation of the SIM-aged martensite variant.
- The characteristic exponents, calculated either from all of the data points, or from the separation of the two types of AE signals (belonging to re-nucleation of the SIM aged variant as well as to reorientations of the competing variants), are the same within the error bars.
- The normalized universal temporal shapes of the avalanches for a fixed area, using the amplitude, A, and $A^{1-\varphi}$ scaling parameters (i.e., the $\frac{U}{A}$ versus $\frac{t}{A^{1-\varphi}}$ plots), were very well scaled together, with φ values obtained from the $E \sim A^{3-\varphi}$ and $S \sim A^{2-\varphi}$ cross-correlation exponents. The tail of this function decays slower by a smaller exponent than the theoretically predicted one, which can be attributed to either the intrinsic absorption of AE signals and/or the overlap of sub-avalanches.

Author Contributions: Conceptualization, D.L.B. and L.D.; methodology, L.D., E.Y.P., Y.I.C. and S.M.K.; software, L.Z.T.; validation, D.L.B., E.Y.P., Y.I.C., L.D. and L.Z.T.; formal analysis, D.L.B., L.D. and L.Z.T.; investigation, L.D. and S.M.K.; data curation, L.D. and L.Z.T.; writing—original draft preparation, D.L.B. and L.D.; writing—review and editing, D.L.B. and L.D.; visualization, L.D.; supervision, D.L.B. and L.D.; All authors have read and agreed to the published version of the manuscript.

Funding: This research was supported by the Ministry of Science and Higher Education of the Russian Federation, project No FSWM-2024-0007.

Data Availability Statement: The original contributions presented in this study are included in the article. Further inquiries can be directed to the corresponding author.

Conflicts of Interest: The authors declare no conflicts of interest.

References

1. Otsuka, K.; Wayman, C.M. *Shape Memory Materials*; Cambridge University Press: Cambridge, UK, 1999.
2. Ren, X.; Otsuka, K. Origin of rubber-like behaviour in metal alloys. *Nature* **1997**, *389/9*, 579–582. [CrossRef]
3. Alaneme, K.K.; Anaele, J.U.; Okotete, E.A. Martensite aging phenomena in Cu-based alloys: Effects of structural transformation, mechanical and shape memory properties: A critical review. *Sci. Afr.* **2021**, *12*, e00760. [CrossRef]
4. Ryklina, E.; Polyakova, K.; AKonopatsky, A.; Teplyakova, C.; Murygin, S.; Andreev, V.; Poletika, T.; Girsova, S.; Komarov, V.; Piskunova, P. Effect of original structure on aging-induced microstructure and transformation behavior of Ni-rich NiTi alloy using various aging modes. *J. Alloys Comp.* **2025**, *1010*, 177859. [CrossRef]
5. Niendorf, T.; Krooβ, P.; Somsen, C.; Eggeler, G.; Chumlyakov, Y.I.; Maier, H.J. Martensite aging—avenue to new high temperature shape memory alloys. *Acta Mater.* **2015**, *89*, 298–304. [CrossRef]
6. Panchenko, E.; Timofeeva, E.; Eftifeeva, A.; Osipovich, S.; Surikov, N.; Chumlyakov, Y.; Gerstein, G.; Maier, H.J. Giant rubber-like behavior induced by martensite aging in Ni51Fe18Ga27Co4 single crystals. *Scr. Mater.* **2019**, *162*, 387–390. [CrossRef]
7. Panchenko, E.Y.; Timofeeva, E.E.; Chumlyakov, Y.I.; Osipovich, K.S.; Tagiltsev, A.I.; Gerstein, G.; Maier, H.J. Compressive shape memory actuation response of stress-induced martensite aged Ni51Fe18Ga27Co4 single crystals. *Mat. Sci. Eng.* **2019**, *A746*, 448–455. [CrossRef]
8. Morito, H.; Fujita, A.; Oikawa, K.; Ishida, K.; Fukamichi, K.; Kainuma, R. Stress assisted magnetic-field-induced strain in Ni-Fe-Ga-Co ferromagnetic shape memory alloys. *Appl. Phys. Lett.* **2007**, *90*, 062505. [CrossRef]
9. Rajput, G.S.; Vora, J.; Prajapati, P.; Chaudhari, R. Areas of recent developments for shape memory alloy: A review. *Mater. Today Proc.* **2022**, *62*, 7194. [CrossRef]
10. Lauhoff, C.; Reul, A.; Langenkämper, D.; Krooß, P.; Somsen, C.; Gutmann, M.J.; Pedersen, B.; Kireeva, I.V.; Chumlyakov, Y.I.; Eggeler, G.; et al. Effects of aging on the stress-induced martensitic transformation and cyclic superelastic properties in Co-Ni-Ga shape memory alloy single crystals under compression. *Acta Mater.* **2022**, *226*, 117623. [CrossRef]
11. Reul, A.; Lauhoff, C.; Krooß, P.; Somsen, C.; Langenkämper, D.; Gutmann, M.J.; Pedersen, B.; Hofmann, M.; Gan, W.M.; Kireeva, I.; et al. On the impact of nanometric γ ′ precipitates on the tensile deformation of superelastic Co49Ni21Ga30. *Acta Mater.*. [CrossRef]
12. Bronstein, E.; Tóth, L.Z.; Daróczi, L.; Beke, D.L.; Talmon, R.; Shilo, D. Tracking Twin Boundary Jerky Motion at Nanometer and Microsecond Scales. *Adv. Funct. Mater.* **2021**, *31*, 2106573. [CrossRef]
13. Perevertov, A.; Sevcik, M.; Heczko, O. Correlation between Acoustic Emission and Stress Evolution during Single Twin Boundary Motion in Ni-Mn-Ga Magnetic Shape Memory Single Crystal. *Phys. Status Solidi. RRL* **2022**, *17*, 2200245. [CrossRef]
14. Weidner, A.; Vinogradov, A.; Vollmer, M.; Krooß, P.; Kriegel, M.J.; Klemm, V.; Chumlyakov, Y.; Niendorf, T.; Biermann, H. In situ characterization of the functional degradation of a [0,1] orientedFe–Mn–Al–Ni single crystal under compression using acoustic emission measurements. *Acta Mater.* **2021**, *220*, 11733. [CrossRef]
15. Setna, J.P.; Dahmen, K.A.; Myers, C.R. Crackling noise. *Nature* **2001**, *410*, 242–250. [CrossRef]
16. LeBlanc, M.; Angheluta, L.; Dahmen, K.; Goldenfeld, N. Universal fluctuations and extreme statistics of avalanches near the depinning transition. *Phys. Rev. E* **2013**, *87*, 022126. [CrossRef]
17. Papanikolaou, S.; Bohn, F.; Sommer, R.L.; Durin, G.; Zapperi, S.S.; Setna, J.P. Universality beyond power laws and the average avalanche shape. *Nat. Phys.* **2011**, *7*, 316–320. [CrossRef]
18. Kamel, S.M.; Samy, N.M.; Tóth, L.Z.; Daróczi, L.; Beke, D.L. Denouement of the Energy-Amplitude and Size-Amplitude Enigma for Acoustic-Emission Investigations of Materials. *Materials* **2022**, *15*, 4556. [CrossRef]
19. Casals, B.; Dahmen, K.A.; Gou, B.; Rooke, S.; Salje, E.K.H. The duration-energy-size enigma for acoustic emission. *Sci. Rep.* **2021**, *11*, 5590. [CrossRef] [PubMed]

20. Spark, G.; Maas, R. Shapes and velocity relaxation of dislocation avalanches in Au and Nb microcrystals. *Acta Mater.* **2018**, *152*, 86–95. [CrossRef]
21. Planes, A.; Manosa, L.; Vives, E. Acoustic emission in martensitic transformations. *J. Alloys Compd.* **2013**, *577*, S699–S704. [CrossRef]
22. Dobrinevski, A.; Le Doussal, P.; Weise, K.J. Avalanche shape and exponents beyond mean-field theory. *Europhys. Lett.* **2015**, *108*, 66002. [CrossRef]
23. Vu, C.-C.; Weiss, J. Asymmetric Damage Avalanche Shape in Quasibrittle Materials and Subavalanche (Aftershock) Clusters. *Phys. Rev. Lett.* **2020**, *125*, 105502. [CrossRef]
24. Antonaglia, J.; Wright, W.J.; Gu, X.; Byer, R.R.; Hufnagel, T.C.; LeBlanc, M.; Uhl, J.T.; Dahmen, K.A. Bulk metallic glasses deform via avalanches. *Phys. Rev. Lett.* **2014**, *112*, 1555501. [CrossRef] [PubMed]
25. Tóth, L.Z.; Bronstein, E.; Daróczi, L.; Shilo, D.; Beke, D.L. Scaling of average avalanche shapes for acoustic emission during jerky motion of single twin boundary in single-crystalline Ni2MnGa. *Materials* **2023**, *16*, 2089. [CrossRef]
26. Kamel, S.M.; Daróczi, L.; Tóth, L.Z.; Panchenko, E.Y.; Chumlyakov, Y.I.; Samy, N.M.; Beke, D.L. Acoustic emission and DSC investigations of anomalous stress-stain curves and burst like shape recovery of Ni49Fe18Ga27Co6 shape memory single crystals. *Intermetallics* **2023**, *159*, 107932. [CrossRef]
27. Clauset, A.; Shalizi, C.R.; Newman, M.E. Power-Law Distributions in Empirical Data. *SIAM Rev.* **2009**, *51*, 661–703. [CrossRef]
28. Beke, D.L.; Bolgar, M.K.; Toth, L.Z.; Daroczi, L. On the asymmetry of the forward and reverse martensitic transformations in shape memory alloys. *J. Alloys Compd.* **2018**, *741*, 106–115. [CrossRef]
29. Otsuka, K.; Ren, X. Mechanism of martensite aging effects and new aspects. *Mater. Sci. Eng.* **2001**, *A312*, 207–218. [CrossRef]
30. Timofeeva, E.; Panchenko, E.Y.; Tokhmetova, A.B.; Eftifeeva, A.B.; Chumlyakov, Y.I.; Volochaev, M.N. The cyclic stability of rubber-like behaviour in stress-induced martensite aged Ni49Fe18Ga27Co6 (at.%) single crystals. *Mat. Lett.* **2021**, *300*, 130207. [CrossRef]
31. Li, J.; Zhang, J.-z.; Zeng, L.-y.; Wang, S.; Song, X.-y.; Chen, N.-l.; Zuo, X.-w.; Rong, Y.-h. Revealing dislocation activity modes 980 during yielding and uniform deformation of low-temperature tempered steel by acoustic emission. *J. Iron Steel Res. Int.* **2024**, *31*, 3022–3036. [CrossRef]
32. Zhao, Y.; Ding, X.; Sun, J.; Salje, E.K.H. Thermal and athermal crackling noise in ferroelastic nanaostuctures. *J. Phys. Condens Matter* **2014**, *26*, 142201. [CrossRef]
33. Nataf, G.; Salje, E.K.H. Avalanches in ferroelectric and coelastic materials: Phase transition, domain switching and propagation. *Feroelectrics* **2020**, *569*, 82–107. [CrossRef]
34. Zreihan, N.; Faran, E.; Vives, E.; Planes, A.; Shilo, D. Coexistence of a well-determined kinetic law and a scale-invariant power law during the same physical process. *Phys. Rev. B* **2018**, *97*, 014103. [CrossRef]
35. Zreihan, N.; Faran, E.; Vives, E.; Planes, A.; Shilo, D. Relations between stress drops and acoustic emission measured during mechanical loading. *Phys. Rev. Mater* **2019**, *3*, 043603. [CrossRef]
36. Daróczi, L.; Piros, E.; Tóth, L.Z.; Beke, D.L. Magnetic field induced random pulse trains of magnetic and acoustic noises in martensitic single-crystal Ni2MnGa. *Phys. Rev. B* **2016**, *96*, 014416. [CrossRef]
37. Chen, Y.; Gou, B.; Yuan, B.; Ding, X.; Sun, J.; Salje, E.K.H. Multiple Avalanche Processes in Acoustic Emission Spectroscopy: Multibranching of the Erenergy-Amplitude Scaling. *Phys. Stat. Sol. B* **2021**, *259*, 2100465. [CrossRef]
38. Chen, Y.; Gou, B.; Ding, X.; Sun, J.; Salje, E.K.H. Fine structure of acoustic emission spectra: How to separate dsiclocation movements and entanglements in 316L stainless steel. *Appl. Phy. Lett.* **2020**, *117*, 262901. [CrossRef]
39. Chen, Y.; Gou, B.; Ding, X.; Sun, J.; Salje, E.K.H. Real-time monitoring dislocations, martensitic transfromations and detwinning in stainless steel: Statistical analysis and mashine learning. *J. Mat. Sci Technol.* **2021**, *92*, 31–39. [CrossRef]
40. Chen, Y.; Gou, B.; Xu, X.; Ding, X.; Sun, J.; Salje, E.K.H. Multibranches of acoustic emission as identifier for deformation mechanisms in additively manufactured 316L stainless steel. *Addit. Manuf.* **2023**, *78*, 103819. [CrossRef]
41. McFaul, L.W.; Wright, W.J.; Sickle, J.; Dahmen, J.A. Force oscillations distort avalanche shapes. *Mater. Res. Lett.* **2019**, *7*, 456–502. [CrossRef]
42. Utsu, T.; Ogata, Y.; Matsu'ura, R.S. The Centenary of the Omori Formula for a Decay Law of Aftershock Activity. *J. Phys. Earth* **1995**, *43*, 1–33. [CrossRef]

metals

MDPI

Article

Decreasing Electrical Resistivity of Ag Film by Low-Temperature Evaporation and Sintering through Azeotrope Application

Sang Hoon Jung [1], Jae Eun Park [1] and Jong-Hyun Lee [1,2,*]

[1] Department of Materials Science and Engineering, Seoul National University of Science and Technology, Seoul 139-743, Republic of Korea

[2] Research Institute for Future Convergence Materials, Seoul National University of Science and Technology, Seoul 139-743, Republic of Korea

* Correspondence: pljh@snut.ac.kr

Abstract: In the temperature-sensitive components, such as perovskite solar cells, large-area electrical connections with high electrical conductivity are also required. To fulfill the requirements, low-temperature evaporation was realized by preparing binder-free pastes with Ag flakes and a solvent mixture, followed by sintering at 140 °C. The mixed solvent was based on viscous α-terpineol with the addition of an appropriate amount of dipropylene glycol methyl ether acetate or diethylene glycol diethyl ether to achieve an azeotrope composition, followed by the addition of a low-molecular-weight hydroxypropyl cellulose to increase the viscosity and thixotropy. During sintering at 140 °C in air for up to 30 min, the paste with 49.5 wt% α-terpineol, 49.5 wt% dipropylene glycol monomethyl ether acetate, and 1 wt% hydroxypropyl cellulose mixture exhibited an excellent electrical conductivity of 7.72×10^{-6} Ω·cm despite the implementation of low-temperature sintering. The excellent processability of the prepared Ag-based pastes at 140 °C demonstrated their potential for novel application areas.

Keywords: binder-free Ag paste; viscous solvent; hydroxypropyl cellulose; azeotrope composition; electrical resistivity; low-temperature evaporation; sintering

1. Introduction

Perovskite solar cells are actively being developed as window-type solar cells owing to their advantages of translucency and color generation [1–13]. Transparent electrode materials with excellent transparency and electrical conductivity should be developed to fabricate high-efficiency window-type perovskite solar cells. A representative example is pattern printing using a Ag particle-based paste, which has excellent oxidation resistance and sinterability [14–18]. However, the efficiency of perovskite materials can decrease when heat-treated at temperatures of 100 °C. As such, electrode materials and printing processes that provide high conductivity and light transmittance at low process temperatures are required [18]. For pastes containing resin, including low-temperature curing resins, such as polydimethylsiloxane (silicone), acrylic, or phenoxy, excellent electrical conductivity is difficult to achieve owing to the formation of percolation networks by contact between Ag particles [19–23]. For example, Fang et al. developed a hybrid silver paste by adding micro- and nano-sized Ag fillers into epoxy and organic vehicles for application in thermally sensitive flexible electronics. They reported an electrical resistivity of 8.1×10^{-5} Ω·cm after sintering at 150 °C for 30 min [24]. In contrast, for resin-free formulations with lower appropriate viscosity, only Ag particles remain with the evaporation of viscous solvents, facilitating sintering between the particles and achieving excellent electrical conductivity [25–31]. For example, Liu et al. mixed spherical and flake silver particles and obtained an electrical resistivity of 3.31×10^{-5} Ω·cm after sintering at 200 °C for

45 min [25]. Also, Zhao et al. presented superior electrical conductivity characteristics, reporting a resistivity of 1.08×10^{-5} Ω·cm after sintering micron-sized silver particles at 140 °C for 30 min [26].

In this study, a low-temperature sintering paste was prepared by mixing Ag particles with a mixture of solvents, which can be removed at 140 °C by evaporation, and a thickener. Flake-shaped Ag microparticles were used as fillers; however, they are less sinterable than nanoparticles, which are more cost-effective and provide better uniform mixing, making them easier to manufacture into a paste. The mixing ratio of the viscous solvents and thickeners, which was considered the main factors for improving the electrical conductivity by enhanced sinterability, was optimized by measuring the electrical resistivity and microstructure of the formed film with different ratios. The electrical conductivity of the film formed in this resin-free paste was directly affected by the sintering degree between the Ag flakes, which significantly changed with evaporation and removal of solvents and thickeners [32]. Most viscous solvents exhibited low evaporation rates at 140 °C, reducing the sinterability between the Ag flakes and ultimately deteriorating the low-temperature processability of the paste. Therefore, in this study, an azeotrope was formed by the optimal combination of two solvent types. Based on the excellent evaporation behavior at 140 °C with the azeotrope formations, rapid sintering between the Ag flakes was induced to form a film with excellent electrical conductivity.

2. Materials and Methods

2.1. Ingredients for Paste

Ag flakes (327077-50G, 99.9%, Sigma-Aldrich Co., Ltd., St. Louis, MO, USA) were used as the fillers, and paste formulation included α-terpineol (97%, Daejung Chemicals & Metals Co., Ltd., Siheung, Republic of Korea) as the main viscous solvent. Dipropylene glycol monomethylether (DPMA, 99%, Thermo Fisher Scientific Inc., Waltham, MA, USA) and diethylene glycol dimethyl ether (DGDE, 99%, Thermo Fisher Scientific Inc.) were used as azeotrope-forming solvents, and hydroxypropyl cellulose (HPC; H0474, Tokyo Chemical Industry Co., Ltd., Tokyo, Japan) was used as both a thickener and a thixotropy agent. The role of the thickener is to increase viscosity with a minimal amount, contributing to the stable formation of electrode patterns. The organic ingredients were summarized in Table 1.

Table 1. Chemical and physical properties of viscous solvents and powders used in the paste preparation.

Solvent	Molecular Formular	Chemical Structure	Molecular Weight (g/mol)	Melting Point (°C)	Boiling Point (°C)	Decomposition Temp. (°C)	Viscosity (mPa·s at 20 °C)
α-Terpineol	$C_{10}H_{18}O$	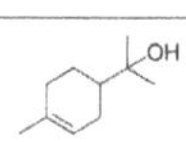	154.25	31	210	-	20
DPMA	$C_{10}H_{18}O_3$	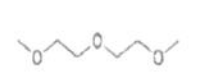	190.23	−23	216	-	4.5
DGDE	$C_6H_{14}O_4$	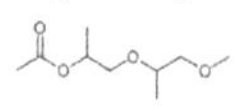	162.23	−64	190	-	0.95
HPC	$(C_6H_{10}O_5)n$ $[(OC_3H_6OH)(OH)_3]_{n-3}$	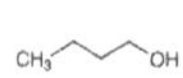	100,000	-	-	130–200	6–10 (2% aqueous solution)

2.2. Paste Preparation and Sintering

The mixed azeotropic solvents were prepared by placing the solvents in a glass bottle at specific ratios and mixing at 2000 rpm for 10 min using a vortex mixer (VM-96M, Jeiotech, Daejeon, Republic of Korea). For the thixotropic formulations, an appropriate HPC amount was added and mixed for an additional hour to prepare the final mixed solvent. Subsequently, the Ag flakes were mixed with the mixed solvents at a weight ratio of 85:15 to prepare a paste. The prepared paste was printed on a glass slide using a stencil mask with a slit volume of 5.0 mm × 5.0 mm × 0.1 mm. The printed films were sintered in a furnace at 140 °C for 10–30 min in air. The average heating rate was 50 °C/min to the temperature of 140 °C. The sintering time was calculated from the insertion point of the sample when the furnace temperature reached 140 °C. After sintering, the samples were removed from the furnace and cooled in air.

2.3. Characterization

The particle sizes of the Ag flakes were measured using a laser particle size analyzer (PSA, LS I3 320, Beckman counter, Brea, CA, USA), and the d_{50} values and average sizes were calculated. The evaporation behavior of the solvents was analyzed through weight measurements using a moisture analyzer (WBA-110M, Daihan Scientific Co., Ltd., Wonju, Republic of Korea). Specifically, 3 g of solvent was placed on the plate of the moisture analyzer and heated to 140 °C. The weight change over time was monitored until the solvent was completely evaporated. The morphologies of the Ag flakes and microstructures of the sintered films after heat treatment were observed using high-resolution field-emission scanning electron microscopy (HR-FE-SEM, SU8010, Hitachi, Tokyo, Japan). Additionally, the amount of residual HPC in the sintered films was confirmed by a carbon content analysis using energy-dispersive X-ray spectroscopy (EDS, Noran System 7, Thermo Fisher Scientific Inc., Waltham, MA, USA). The thermal behaviors of the prepared pastes were confirmed using thermogravimetry differential thermal analysis (TG-DTA, DTG-60, Shimadzu, Kyoto, Japan) by heating to 140 °C at a rate of 10 °C/min and maintaining the temperature for 60 min. The electrical properties of the sintered films were evaluated by electrical resistance measurements ten times using a four-point probe connected to a source meter (2400, Keithley, Cleveland, OH, USA). The electrical resistivity was calculated to extract the average value and standard deviation.

3. Results and Discussion

Figure 1 shows the SEM images of the Ag flakes, with the average size of 5.60 μm and agglomeration, primarily around smaller flakes. Figure 2 shows the particle-size distribution with the measured d_{10}, d_{50}, and d_{90} values of 1.79, 4.05, and 10.37 μm, respectively, indicating a bimodal size distribution with flakes in the tens of micrometers range.

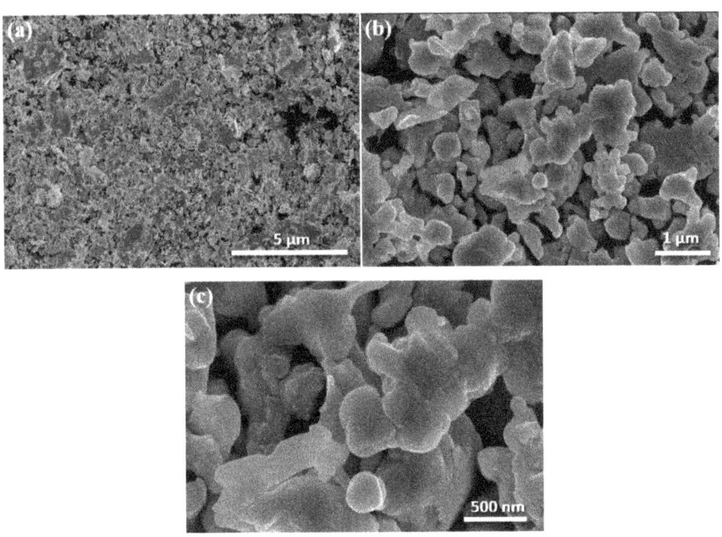

Figure 1. SEM images of the Ag flakes at different magnifications: (**a**) 2000, (**b**) 10,000, and (**c**) 50,000 times.

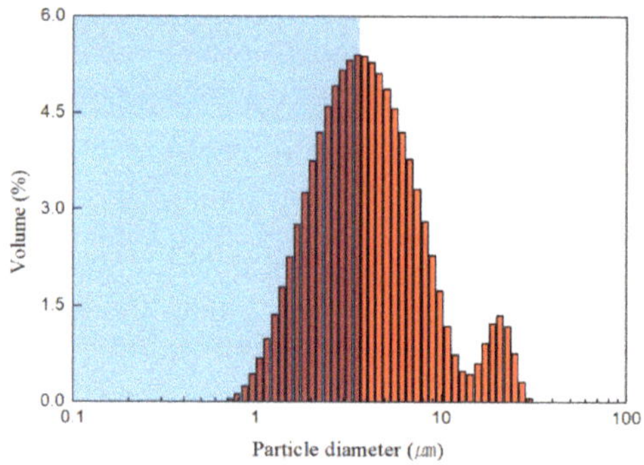

Size (μm)	Min.	d10	d50	d90	Max.	Mean
	0.60	1.79	4.05	10.37	36.24	5.60

Figure 2. Laser PSA analysis result of the Ag flakes.

Figure 3 shows the evaporation behavior of the main viscous solvent, α-terpineol, and two azeotrope-forming solvents (DPMA and DGDE) at 140 °C over time. All three solvents evaporated completely in 20 min; however, their evaporation rates varied significantly. Specifically, the evaporation rate was as follows: DGDE > DPMA > α-terpineol. This indicates the inverse proportional relationship between the viscosity and evaporation rate.

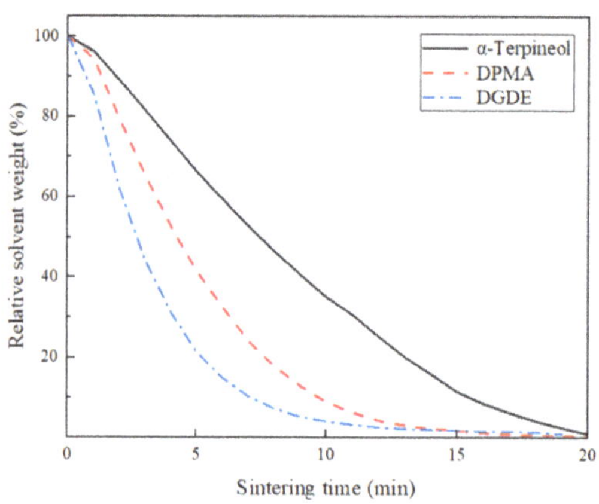

Figure 3. Weight loss of α-terpineol, DPMA, and DGDE over time at 140 °C.

The evaporation behavior changes with the amount of azeotrope-forming solvent added to induce rapid evaporation of α-terpineol, as presented in Figure 4. The most drastic results were obtained when DPMA was mixed with α-terpineol (Figure 4a), with the mixed solvent exhibiting a higher evaporation rate than pure DPMA, achieving the rapid evaporation by azeotrope formation. At a 50:50 weight ratio, evaporation was completed in approximately 11.7 min, with the highest evaporation rate. In contrast, the evaporation rate with DGDE mixed with α-terpineol was between that of pure α-terpineol and pure DPMA when one solvent was in excess (Figure 4b). However, the evaporation rate at a 50:50 mixing ratio is higher than that of pure DGDE, achieving the desired result. Therefore, mixtures of α-terpineol and DPMA and α-terpineol and DGDE at compositions of 50:50 were considered as the basic compositions of the mixed solvent in subsequent studies for preparing an optimal paste. When DPMA and DGDE were mixed without α-terpineol (Figure 4c), the attempted mixing ratios had a lower evaporation rate than pure DGDE, indicating the absence of azeotrope formation. Based on these results, eight mixed solvent compositions with a small amount of thickener are listed in Table 2. HPC, which was used as a thickener, significantly improved the viscosity and thixotropy of the paste, even with small additions (1, 2, and 3 wt%), leading to significant improvements in the printability of the paste. This is an important factor that significantly affected the sinterability between the Ag flakes and the electrical resistivity of the sintered film.

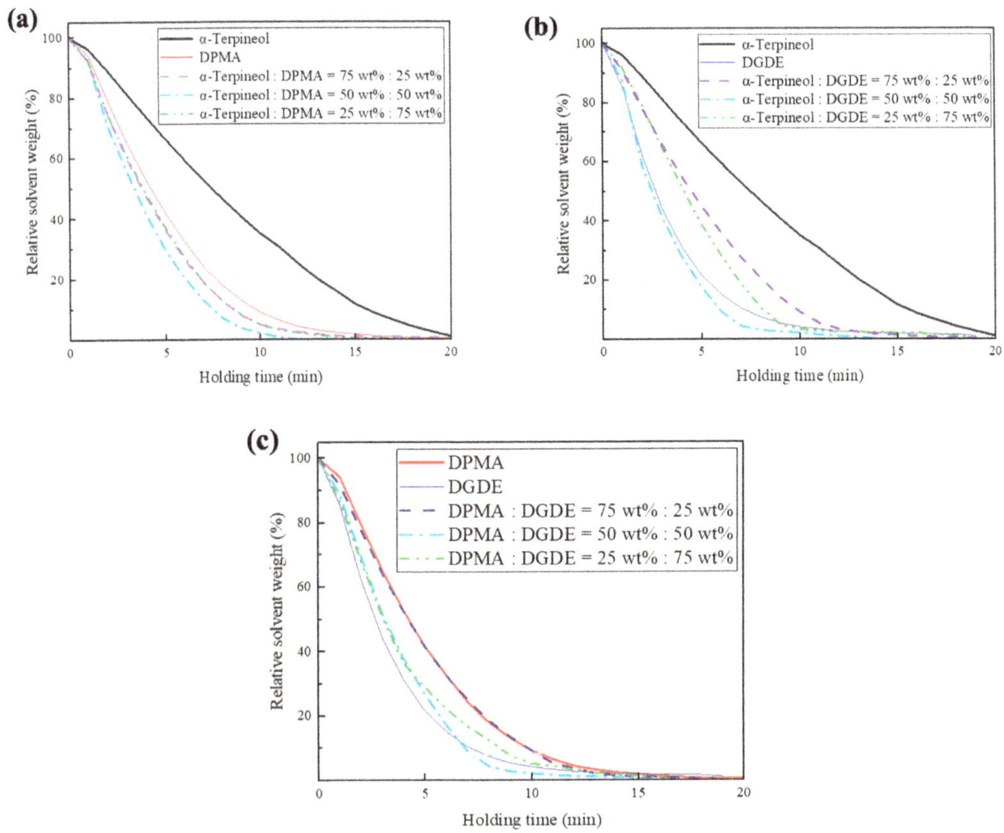

Figure 4. Weight-loss graphs of nine mixed solvents over time at 140 °C: (**a**) α-terpineol and DPMA, (**b**) α-terpineol and DGDE, and (**c**) DPMA and DGDE.

Table 2. Compositions of the investigated solvent mixtures for the paste formulation.

Solvent	Solvent 1	Solvent 2	HPC
#1	a-Terpineol (50 wt%)	DPMA (50 wt%)	0 wt%
#2	a-Terpineol (49.5 wt%)	DPMA (49.5 wt%)	1 wt%
#3	a-Terpineol (49 wt%)	DPMA (49 wt%)	2 wt%
#4	a-Terpineol (48.5 wt%)	DPMA (48.5 wt%)	3 wt%
#5	a-Terpineol (50 wt%)	DGDE (50 wt%)	0 wt%
#6	a-Terpineol (49.5 wt%)	DGDE (49.5 wt%)	1 wt%
#7	a-Terpineol (49 wt%)	DGDE (49 wt%)	2 wt%
#8	a-Terpineol (48.5 wt%)	DGDE (48.5 wt%)	3 wt%

Figure 5 shows the evaporation of the eight mixed solvents, including those with 1–3 wt% HPC thickener at 140 °C over time. In the α-terpineol/DPMA mixed solvents (Figure 5a), the evaporation rate decreased with increasing HPC content. With 1 wt% HPC, the mixed solvents are almost completely evaporated in 12.5 min. Similarly, the evaporation rate in the α-terpineol/DGDE mixed solvents (Figure 5b) decreases with increasing HPC content, whereas that with 1 wt% HPC decreases minimally, completely evaporating the solvent in 8 min. Therefore, under identical processing conditions, compositions with 1 wt% HPC are identified as the most effective for inducing rapid sintering in high-viscosity α-terpineol/DPMA mixed solvents and low-viscosity α-terpineol/DGDE mixed solvents.

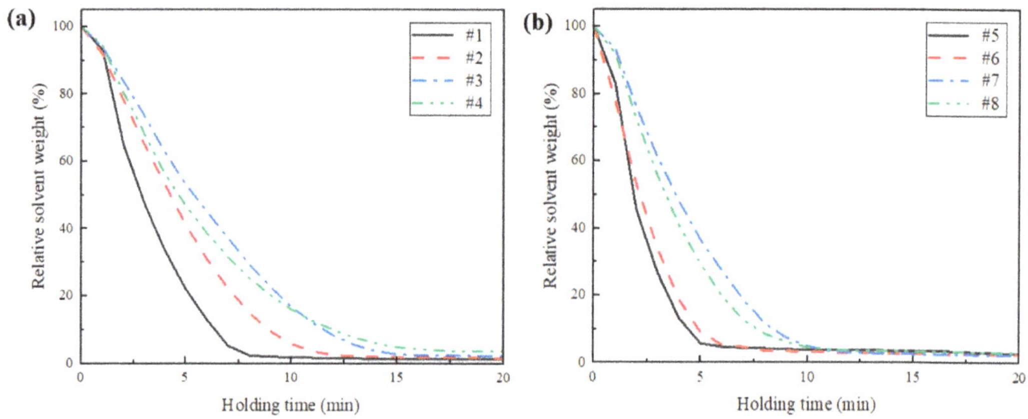

Figure 5. Weight-loss graphs of eight mixed solvents with the HPC thickener over time at 140 °C: (**a**) α-terpineol/DPMA mixed solvents and (**b**) α-terpineol/DGDE mixed solvents.

The TG-DTA results of the pastes prepared with various mixed solvents under dynamic and isothermal mixed heating conditions, similar to the actual sintering temperature profile, are shown in Figure 6. In the DPMA-containing specimens (Figure 6a), an endothermic reaction began during the heating onset. Significant weight loss by solvent evaporation is observed at temperatures above 70 °C. Endothermic peaks are observed in the range of 130.0–139.1 °C, and continuous weight loss is observed until reaching 140 °C. However, in the sample with 1 wt% HPC (#2), a small exothermic behavior at 103 °C and a delay in weight loss are observed, which could be attributed to the sintering of Ag flakes. Moreover, the delay in weight loss can be ascribed to the blockage of the outgassing pathways by sintering. Meanwhile, in the DGDE-containing specimens (Figure 6b), weight loss due to solvent evaporation is observed, starting at a notably lower temperature of 40 °C in the sample with 1 wt% HPC (#6), and completing before reaching 140 °C. The endothermic peak ascribed to solvent evaporation is observed at 126.4 °C, which is considerably lower temperature than the other mixed solvents owing to the formation of an azeotrope with an appropriate HPC amount, thereby improving the evaporation behavior and positively affecting the sintering of Ag flakes. For sample #6, a small exothermic peak is observed at approximately 140 °C after solvent evaporation, which can be ascribed to the sintering of Ag flakes.

The printability of the eight pastes is shown in Figure 7. For the pastes without HPC (#1 and #5), the low viscosity and thixotropy collapsed the printed patterns. However, these issues are mitigated as the HPC content is increased. In particular, in the DPMA-containing pastes, which were predicted to have higher viscosity, excellent printability with almost no pattern collapse is observed in the composition with 2 wt% HPC. Except for #1 and #5, the average thickness of the printed patterns was measured to be 92.5 (±2.1) μm.

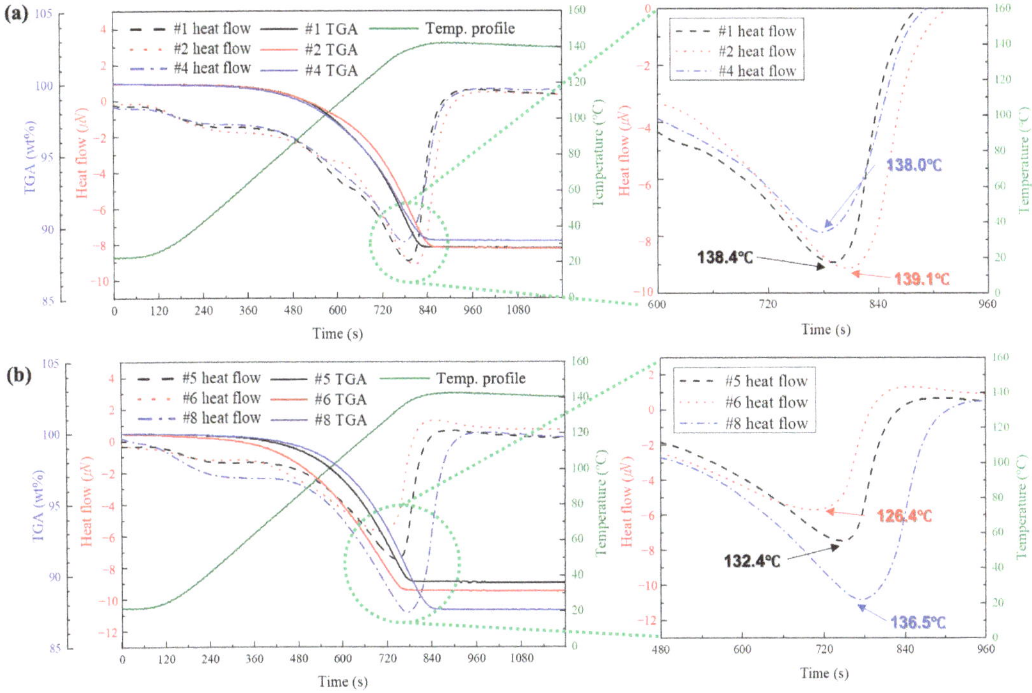

Figure 6. TG-DTA curves of the Ag pastes with (**a**) α-terpineol/DPMA(/HPC) and (**b**) DGDE/α-Terpineol(/HPC) mixtures.

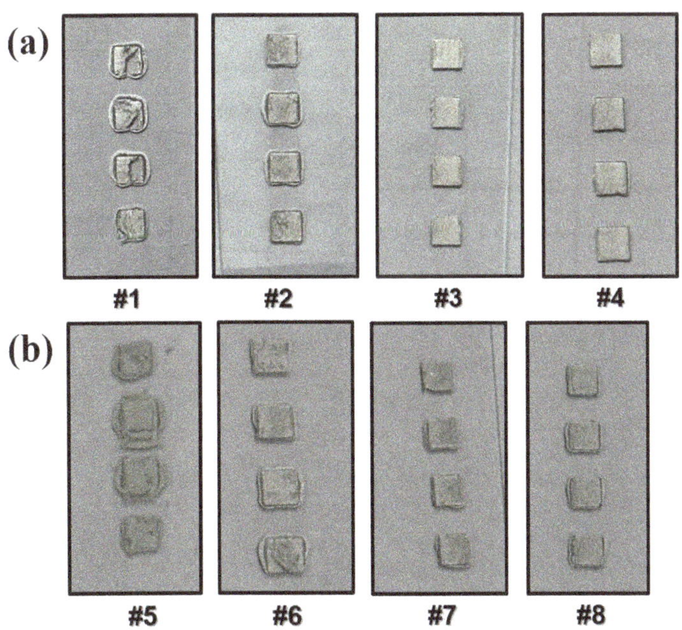

Figure 7. Images of screen-printed films using Ag pastes prepared with different mixed solvents and HPC contents: (**a**) α-terpineol/DPMA(/HPC) and (**b**) DGDE/α-Terpineol(/HPC) mixtures.

Figure 8 shows the electrical resistivities and sheet resistances of films from the printed patterns after sintering at 140 °C with respect to the Ag paste type and sintering time. For the specimens containing DPMA (Figure 8a), the resistivity of the pastes gradually decreases with increasing sintering time. Compared to the paste without HPC (#1), HPC addition significantly decreased the resistivity after sintering for 20 min. After sintering for 30 min, the resistivity greatly decreased for pastes #1 and #2, whereas a minimal decrease is observed in pastes #3 and #4, which had a higher HPC content. Consequently, paste #2 with 1 wt% HPC has the lowest resistivity of 7.72×10^{-6} Ω·cm obtained in 30 min at 140 °C. Therefore, when compared to previous similar results [14–26], the achieved resistivity is evaluated as an excellent value. This suggests that HPC addition improved the contact between the Ag flakes owing to the improved printability, providing positive results after up to 20 min of sintering. However, in pastes #3 and #4 with higher HPC contents, the residual HPC is expected to limit the sintering of the flakes after 20 min of sintering. In contrast, paste #2 with an appropriate HPC amount continued to exhibit excellent sinterability owing to improved contact between Ag flakes without the hindrance of residual HPC, as observed from the exothermic results obtained starting at 103 °C in Figure 6a.

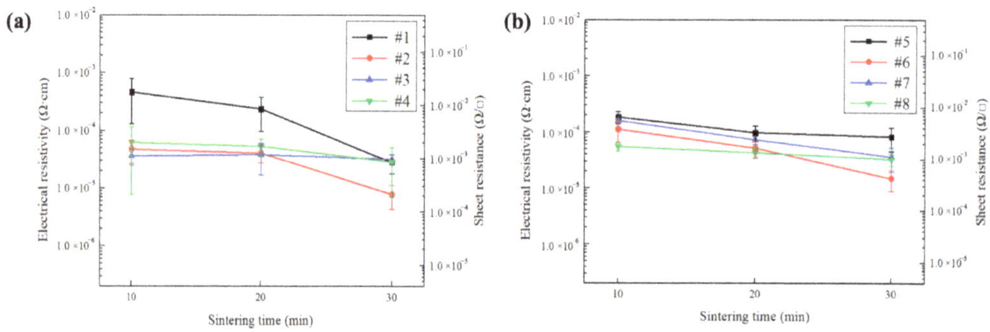

Figure 8. Electrical resistivities and sheet resistances of the films sintered at 140 °C with different HPC contents and sintering times: (**a**) α-terpineol/DPMA mixed solvents and (**b**) α-terpineol/DGDE mixed solvents.

For the specimens with DGDE (Figure 8b), paste #5 inherently provides a lower resistivity, which can be considered an early sintering start effect by previously observed high evaporation rates. All pastes exhibit a gradual decrease in resistivity with increasing sintering time, and the HPC addition tends to decrease the resistivity. However, minimal resistivity reduction due to HPC addition is noted owing to its inherently low viscosity. Similarly, paste #6 with 1 wt% HPC has a low resistivity of 1.47×10^{-5} Ω·cm after sintering for 30 min.

The HPC is an organic compound including C; thus, the residual HPC amount in the sintered film can be compared by the C content. The C content in the film maintained for 30 min was measured by EDS to confirm that the residual amount in the sintered film increases with the HPC content during heating at 140 °C. The results are shown in Figure 9. The measured C content increased with the HPC content in the specimens with DPMA and DGDE, indicating that the initial HPC content in the paste is proportional to the residual HPC content in the sintered film at 140 °C. Therefore, the insufficient resistance values observed in the pastes containing 2 or 3 wt% HPC after 30 min of sintering are attributed to the residual HPC content.

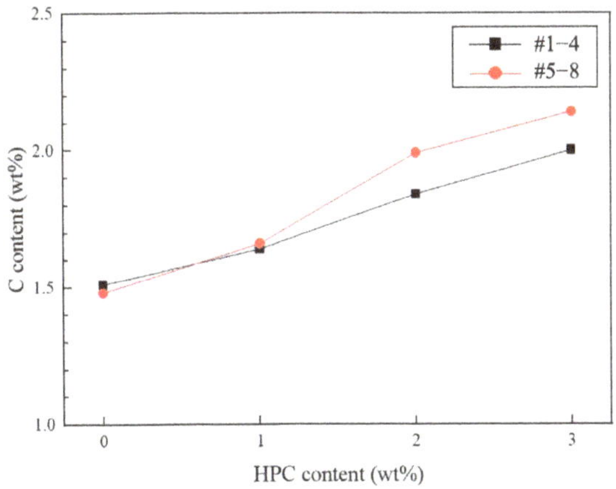

Figure 9. C contents measured in the films based on paste samples with different HPC contents after sintering for 30 min at 140 °C in air, as obtained by EDS.

Figure 10 shows the surface microstructures of the films formed by the pastes with DPMA and varying HPC contents and sintering times at 140 °C. The microstructure of the film sintered for 30 min using paste #2 has the widest sintered area and best connectivity between the flakes compared to the microstructures under all other conditions (note the larger grain indicated by the arrow). Such microstructure corresponds well to the lowest electrical resistivity (7.72×10^{-6} Ω·cm) measured for the paste #2 after 30 min of sintering, as shown in Figure 8a. Figure 11 also shows that paste #6 sintered for 30 min exhibits the largest sintered area with well-established interconnections between the flakes compared with pastes #5, #7, and #8.

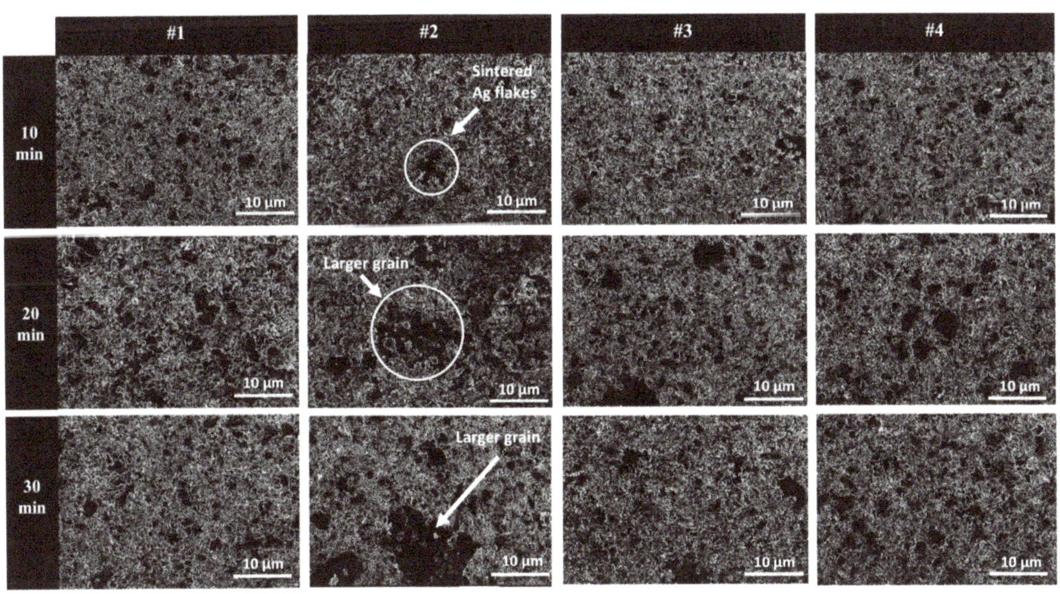

Figure 10. Surface SEM images of the films sintered at 140 °C in air using α-terpineol/DPMA-based pastes with different HPC contents and sintering times.

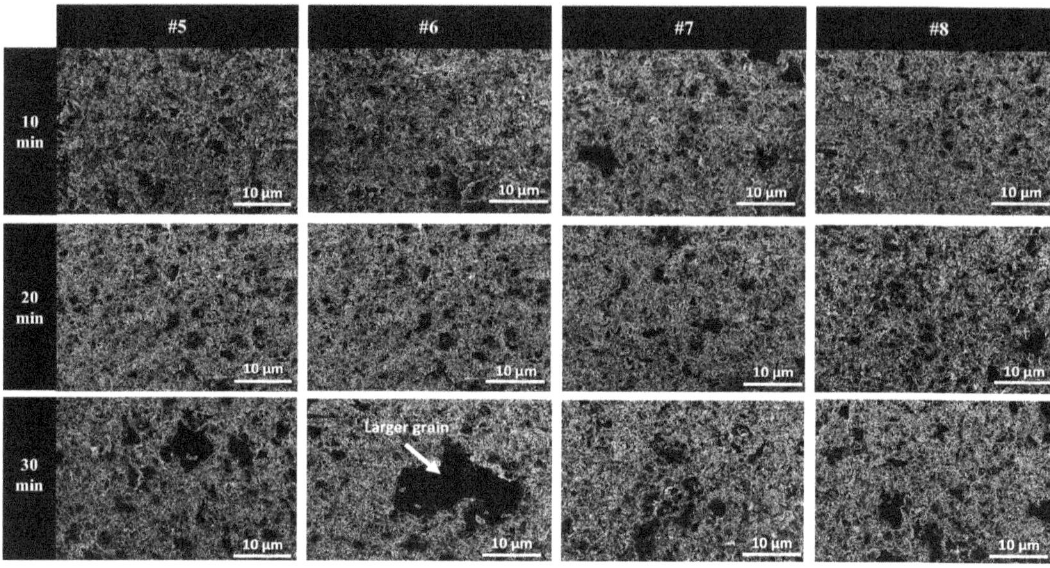

Figure 11. Surface SEM images of the films sintered at 140 °C in air using α-terpineol/DGDE-based pastes with different HPC contents and sintering times.

4. Conclusions

A resin-free low-temperature evaporative solvent-based paste formulation was implemented to form a film with low electrical resistivity through rapid sintering between micron-sized Ag flakes at a low temperature of 140 °C. Specifically, α-terpineol was used as the main viscous solvent, with DGDE or DPMA added as the azeotrope-forming solvents to greatly improve the evaporation rate. Finally, an HPC thickener was added to increase the viscosity and thixotropy. The paste containing 49.5 wt% α-terpineol, 49.5 wt% DPMA, and 1 wt% HPC exhibited improved contacts between the Ag flakes and overall percolation connectivity due to its good printability and early sintering owing to the improved evaporation behavior, resulting in a rapid decrease in the electrical resistivity even at 140 °C. Specifically, the film using this paste and sintered for 30 min exhibited the lowest resistivity of 7.72×10^{-6} Ω·cm, with the early exothermic behavior starting at 103 °C owing to sintering. The microstructure of the formed film exhibited the widest sintered area and best connectivity between flakes. Although increasing the HPC content to 3 or 4 wt% further improved the printability, the resistivity values slightly increased after sintering for 30 min owing to the increased residual HPC. The paste with 49.5 wt% α-terpineol, 49.5 wt% DGDE, and 1 wt% HPC demonstrated reduced printability due to the decreased viscosity, whereas the maximum evaporation rate obtained with 1 wt% HPC resulted in the highest resistivity of 1.47×10^{-5} Ω·cm after sintering for 30 min.

Author Contributions: Conceptualization, J.-H.L.; methodology, S.H.J. and J.-H.L.; validation, S.H.J. and J.E.P.; formal analysis, S.H.J., J.E.P. and J.-H.L.; investigation, S.H.J., J.E.P. and J.-H.L.; resources, J.-H.L.; data curation, S.H.J. and J.E.P.; writing—original draft preparation, S.H.J. and J.-H.L.; writing—review and editing, J.-H.L.; visualization, S.H.J. and J.E.P.; supervision, J.-H.L.; project administration, J.-H.L.; funding acquisition, J.-H.L. All authors have read and agreed to the published version of the manuscript.

Funding: This study was supported by Korea Electric Power Corporation. (Grant number: R23XO05-02).

Data Availability Statement: The original contributions presented in the study are included in the article, further inquiries can be directed to the corresponding author.

Conflicts of Interest: The authors declare no conflicts of interest. The funders had no role in the design of the study, in the collection, analyses, or interpretation of data, in the writing of the manuscript, or in the decision to publish the results.

References

1. Gao, L.; Chao, L.; Hou, M.; Liang, J.; Chen, Y.; Yu, H.; Huang, W. Flexible, transparent nanocellulose paper-based perovskite solar cells. *npj Flex. Electron.* **2019**, *3*, 1–8. [CrossRef]
2. Lee, K.-T.; Jang, J.-Y.; Park, S.J.; Ok, S.A.; Park, H.J. Incident-angle-controlled semitransparent colored perovskite solar cells with improved efficiency exploiting a multilayer dielectric mirror. *Nanoscale* **2017**, *9*, 13983–13989. [CrossRef] [PubMed]
3. Cui, D.; Yang, Z.; Yang, D.; Ren, X.; Liu, Y.; Wei, Q.; Fan, H.; Zeng, J.; Liu, S. Color-tuned perovskite films prepared for efficient solar cell applications. *J. Phys. Chem. C* **2016**, *120*, 42–47. [CrossRef]
4. Schlisske, S.; Mathies, F.; Busko, D.; Strobel, N.; Rödlmeier, T.; Richards, B.; Lemmer, U.; Paetzold, U.; Hernández-Sosa, G.; Klampaftis, E. Design and color flexibility for inkjet-printed perovskite photovoltaics. *ACS Appl. Energy Mater.* **2019**, *2*, 764–769. [CrossRef]
5. Quiroz, C.O.R.; Bronnbauer, C.; Levchuk, I.; Hou, Y.; Brabec, C.J.; Forberich, K. Coloring semitransparent perovskite solar cells via dielectric mirrors. *ACS Nano* **2016**, *10*, 5104–5112. [CrossRef]
6. Bati, A.S.R.; Zhong, Y.L.; Burn, P.L.; Nazeeruddin, M.K.; Shaw, P.E.; Batmunkh, M. Next-generation applications for integrated perovskite solar cells. *Commun. Mater.* **2023**, *4*, 2. [CrossRef]
7. Bing, J.; Granados Caro, L.; Talathi, H.P.; Chang, N.L.; Mckenzie, D.R.; Ho-Baillie, A.W.Y. Perovskite solar cells for building integrated photovoltaics—Glazing applications. *Joule* **2022**, *6*, 1446–1474. [CrossRef]
8. Zhu, Y.; Shu, L.; Fan, Z. Recent Progress on Semi-transparent Perovskite Solar Cell for Building-integrated Photovoltaics. *Chem. Res. Chin. Univ.* **2020**, *36*, 366–376. [CrossRef]
9. Koh, T.M.; Wang, H.; Ng, Y.F.; Bruno, A.; Mhaisalkar, S.; Mathews, N. Halide Perovskite Solar Cells for Building Integrated Photovoltaics: Transforming Building Façades into Power Generators. *Adv. Mater.* **2022**, *34*, 2104661. [CrossRef]
10. Wang, H.; Li, J.; Dewi, H.A.; Mathews, N.; Mhaisalkar, S.; Bruno, A. Colorful Perovskite Solar Cells: Progress, Strategies, and Potentials. *J. Phys. Chem. Lett.* **2021**, *12*, 1321–1329. [CrossRef]
11. Mujahid, M.; Chen, C.; Zhang, J.; Li, C.; Duan, Y. Recent advances in semitransparent perovskite solar cells. *InfoMat* **2021**, *3*, 101–124. [CrossRef]
12. Nazir, G.; Lee, S.-Y.; Lee, J.-H.; Rehman, A.; Lee, J.-K.; Seok, S.I.; Park, S.-J. Stabilization of Perovskite Solar Cells: Recent Developments and Future Perspectives. *Adv. Mater.* **2022**, *34*, 2204380. [CrossRef] [PubMed]
13. Rakocevic, L.; Gehlhaar, R.; Jaysankar, M.; Song, W.; Aernouts, T.; Fledderus, H.; Poortmans, J. Translucent, color-neutral and efficient perovskite thin film solar modules. *J. Mater. Chem. C* **2018**, *6*, 3034–3041. [CrossRef]
14. Yang, D.; Huang, Y.; Tian, Y. Microstructure of Ag nano paste joint and its influence on reliability. *Crystals* **2021**, *11*, 1537. [CrossRef]
15. Yang, H. Study on the preparation process and sintering performance of doped nano-silver paste. *Rev. Adv. Mater. Sci.* **2022**, *61*, 969–976. [CrossRef]
16. Ma, L.; Wang, Y.; Jia, Q.; Zhang, H.; Wang, Y.; Li, D.; Zou, G.; Guo, F. Low-temperature-sintered nano-Ag film for power electronics packaging. *J. Electron. Mater.* **2024**, *53*, 228–237. [CrossRef]
17. Chen, C.; Choe, C.; Kim, D.; Zhang, Z.; Long, X.; Zhou, Z.; Wu, F.; Suganuma, K. Effect of oxygen on microstructural coarsening behaviors and mechanical properties of Ag sinter paste during high-temperature storage from macro to micro. *J. Alloys Compd.* **2020**, *834*, 155173. [CrossRef]
18. Yan, J. A Review of Sintering-Bonding Technology Using Ag Nanoparticles for Electronic Packaging. *Nanomaterials* **2021**, *11*, 927. [CrossRef]
19. Divitini, G.; Cacovich, S.; Matteocci, F.; Cinà, L.; Di Carlo, A.; Ducati, C. In situ observation of heat-induced degradation of perovskite solar cells. *Nat. Energy* **2016**, *1*, 15012. [CrossRef]
20. Sasaki, K.; Mizumura, N. Development of Low-temperature Sintering Nano-Ag Pastes Using Lowering Modulus Technologies. In Proceedings of the IEEE 17th International Conference on Nanotechnology (IEEE-NANO), Pittsburgh, PA, USA, 25–28 July 2017.
21. Jung, K.-H.; Min, K.D.; Lee, C.-J.; Park, B.-G.; Jung, H.; Koo, J.-M.; Lee, B.; Jung, S.-B. Effect of epoxy content in Ag nanoparticle paste on the bonding strength of MLCC packages. *Appl. Surf. Sci.* **2019**, *495*, 143487. [CrossRef]
22. Ho, L.N.; Wu, T.F.; Nishikawa, H. Properties of Phenolic-Based Ag-Filled Conductive Adhesive Affected by Different Coupling Agents. *J. Adhes.* **2013**, *89*, 847–858. [CrossRef]
23. Wu, H.; Chiang, S.; Han, W.; Tang, Y.; Kang, F.; Yang, C. Surface iodination: A simple and efficient protocol to improve the isotropically thermal conductivity of silver-epoxy pastes. *Compos. Sci. Technol.* **2014**, *99*, 109–116. [CrossRef]
24. Fang, S.; Zhang, Y.; Huang, C.; Liu, P.; Zhang, C.; Liu, Y.; Li, S. Hybrid silver pastes with synergistic effect of multi-scale silver fillers and the application in flexible circuits. *J. Mater. Sci. Mater. Electron.* **2021**, *32*, 13777–13786. [CrossRef]
25. Zhao, L.; Lin, Z.; Liu, Y.; Shi, Z.; Yu, L.; Li, Z. Parametric study on conductive patterns by low-temperature sintering of micron silver ink. *J. Mater. Sci. Mater. Electron.* **2023**, *34*, 653–659. [CrossRef]

26. Liu, X.; Wu, S.; Chen, B.; Ma, Y.; Huang, Y.; Tang, S.; Liu, W. Tuning the electrical resistivity of conductive silver paste prepared by blending multi-morphologies and micro-nanometers silver powder. *J. Mater. Sci. Mater. Electron.* **2021**, *32*, 13777–13786. [CrossRef]
27. Noh, S.; Choe, C.; Chen, C.; Zhang, H.; Suganuma, K. Printed wire interconnection using Ag sinter paste for wide band gap power semiconductors. *J. Mater. Sci. Mater. Electron.* **2018**, *29*, 15223–15232. [CrossRef]
28. Zhang, H.; Nagao, S.; Suganuma, K.; Albrecht, H.-J.; Wilke, K. Thermostable Ag die-attach structure for high-temperature power devices. *J. Mater. Sci. Mater. Electron.* **2016**, *27*, 1337–1345. [CrossRef]
29. Yang, F.; Hu, B.; Peng, Y.; Hang, C.; Chen, H.; Lee, C.; Wei, J.; Li, M. Ag microflake-reinforced nano-Ag paste with high mechanical reliability for high-temperature applications. *J. Mater. Sci. Mater. Electron.* **2019**, *30*, 5526–5532. [CrossRef]
30. Martin, H.A.; Hu, D.; Liu, X.; Poelma, R.H.; Smits, E.C.P.; Van Driel, W.D.; Zhang, G. Prognostic Monitoring of Power QFN Packages with Silver Sintered Die-Attach Materials. *IEEE Trans. Compon. Packag. Manuf. Technol.* **2024**. [CrossRef]
31. Wang, M.; Mei, Y.-H.; Jin, J.; Chen, S.; Li, X.; Lu, G.-Q. Pressureless Sintered-Silver Die-Attach at 180 °C for Power Electronics Packaging. *IEEE Trans. Power Electron.* **2021**, *36*, 12141–12145. [CrossRef]
32. Yao, S.; Xing, J.; Zhang, J.; Xiong, S.; Yang, Y.; Yuan, X.; Li, H.; Tong, H. Microscopic investigation on sintering mechanism of electronic silver paste and its effect on electrical conductivity of sintered electrodes. *J. Mater. Sci. Mater. Electron.* **2018**, *29*, 18540–18546. [CrossRef]

metals MDPI

Article

Improvement in the Hydrogen Storage Properties of MgH_2 by Adding $NaAlH_4$

Young-Jun Kwak [1,2], Myoung-Youp Song [1,2,*] and Ki-Tae Lee [1,2,3]

1 Division of Advanced Materials Engineering, Jeonbuk National University, 567 Baekje-daero, Deokjin-gu, Jeonju 54896, Republic of Korea; twistking18@nate.com (Y.-J.K.); ktlee71@jbnu.ac.kr (K.-T.L.)
2 Hydrogen & Fuel Cell Research Center, Engineering Research Institute, Jeonbuk National University, 567 Baekje-daero, Deokjin-gu, Jeonju 54896, Republic of Korea
3 Department of Energy Storage/Conversion Engineering of Graduate School (BK21 FOUR), Jeonbuk National University, 567 Baekje-daero, Deokjin-gu, Jeonju 54896, Republic of Korea
* Correspondence: songmy@jbnu.ac.kr; Tel.: +82-10-3260-2379

Abstract: Milled MgH_2, MgH_2-10$NaAlH_4$, MgH_2-30$NaAlH_4$, MgH_2-50$NaAlH_4$, and MgH_2-2Ni-10$NaAlH_4$ samples were prepared by milling in a planetary ball mill in hydrogen atmosphere (reactive mechanical milling, RMM). Decomposition temperatures of milled MgH_2, $NaAlH_4$, MgH_2-10$NaAlH_4$, and MgH_2-30$NaAlH_4$ were examined in a Sieverts-type hydrogen absorption and release apparatus, in which the hydrogen pressures were kept nearly constant during hydrogen absorption or release. As the content of $NaAlH_4$ in the sample increased, the temperature at the highest peak in the ratio of increase in released hydrogen quantity to increase in temperature versus temperature curve decreased. Hydriding in 12 bar hydrogen and dehydriding in 1.0 bar hydrogen at 593 K of MgH_2-30$NaAlH_4$ are performed by the reversible reactions $MgH_2 \Leftrightarrow Mg + H_2$ and $17MgH_2 + 12Al \Leftrightarrow Mg_{17}Al_{12} + 17H_2$. MgH_2-30$NaAlH_4$ was the best Mg-based composite among Mg-based alloys in which an oxide, a halide, a fluoride, or a complex hydride was added, with a high hydrogen absorption rate for 2.5 min (2.20 wt% H/min) and a large effective hydrogen storage capacity (7.42 wt% H).

Keywords: hydrogen storage; MgH_2; $NaAlH_4$; decomposition temperature; hydrogen absorption and release rates

1. Introduction

Clean alternative energies have drawn interest in preventing air pollution and climate change. One of the clean alternative energies is hydrogen energy, which is the energy produced by the reaction of hydrogen with oxygen, producing water as a by-product. The problems to be solved for applying hydrogen energy to practical use are hydrogen production and storage.

Hydrogen is usually stored in a gaseous state under high pressure and in a cryogenic liquid state. Storing gaseous hydrogen has disadvantages such as safety issues, high cost, and hydrogen embrittlement of storage tank materials. Storage of hydrogen in a cryogenic liquid state has drawbacks such as thermal losses in the case of an open system, safety, and cost of liquefaction.

Another method to store hydrogen is solid-state hydrogen storage using materials such as metal hydrides. Metal hydrides have advantages such as high gravimetric and volumetric storage capacities and safety, as metal hydrides can absorb and release hydrogen at relatively low pressures. Hydrogen is bound by chemical or physical forces in hydrogen storage based on solid-state materials. The technique of storing hydrogen in a solid state has become very attractive and is the subject of studies by many researchers. We are interested in synthesizing metal hydrides based on a metal, magnesium (Mg). Mg-based hydrides have relatively high reaction rates and high hydrogen storage capacities.

Complex hydrides such as $LiBH_4$ and $NaAlH_4$ have been studied by many researchers because they have high theoretical hydrogen storage capacities [1–10]. Many works were performed to improve the hydriding and dehydriding kinetics of Mg [11–16]. Researchers were interested in improving the hydrogen storage properties of MgH_2 by adding $NaAlH_4$ with a high hydrogen storage capacity [17–21]. Ali and Ismail [17] reviewed the hydrogen storage properties of the Mg–Na–Al system. The complex hydride $NaAlH_4$ releases hydrogen via three-step reactions:

$$\text{At } 458 - 503\text{ K,} \quad 3NaAlH_4 \rightarrow Na_3AlH_6 + 2Al + 3H_2 \tag{1}$$

$$\text{At } 533\text{ K,} \quad Na_3AlH_6 \rightarrow 3NaH + Al + (3/2)H_2 \tag{2}$$

$$\text{At } 708\text{ K,} \quad 3NaH \rightarrow 3Na + (3/2)H_2 \tag{3}$$

Ali and Ismail [17] reported that the addition of $NaAlH_4$ could destabilize the MgH_2 effectively and the hydrogen storage properties of $NaAlH_4$ could also be improved by adding MgH_2. The MgH_2-$NaAlH_4$ system exhibited much better dehydriding properties than unary MgH_2 and $NaAlH_4$.

Ismail et al. [20] reported that the following reaction takes place within the temperature range from 443 to 485 K:

$$NaAlH_4 + MgH_2 \rightarrow NaMgH_3 + Al + (3/2)H_2 \tag{4}$$

A mixing decomposition of the reaction of MgH_2 with Al (4) and the decomposition of the excessive MgH_2 (5) occurs reversibly between 553 K and 603 K [20].

$$17MgH_2 + 12Al \Leftrightarrow Mg_{17}Al_{12} + 17H_2 \tag{5}$$

$$MgH_2 \Leftrightarrow Mg + H_2 \tag{6}$$

They also reported that $NaMgH_3$ decomposes between 603 K and 633 K by the following reversible reaction:

$$NaMgH_3 \Leftrightarrow NaH + Mg + H_2 \tag{7}$$

NaH decomposes between 633 K and 648 K by the following reversible reaction [20]:

$$NaH \Leftrightarrow Na + (\frac{1}{2})H_2. \tag{8}$$

In the present work, samples with various compositions were prepared and many phenomena were examined such as decomposition temperatures, phase formation, hydrogen absorption and release rates, and hydrogen storage capacity. In addition, the hydrogen storage properties of the prepared samples were compared with those of other Mg-based alloys. In particular, the decomposition temperatures were measured in the Sieverts-type hydrogen absorption and release apparatus, in which the samples were hydrided and dehydrided, by heating the samples at a constant heating rate and simultaneously measuring the released hydrogen quantity. In other researchers' work, the decomposition temperatures were usually measured in thermal analysis instruments such as thermal gravimetric analysis (TGA) and differential scanning calorimetry (DSC). Milled MgH_2, MgH_2-10$NaAlH_4$ (with a composition of 90 wt% MgH_2 + 10 wt% $NaAlH_4$), MgH_2-30$NaAlH_4$ (70 wt% MgH_2 + 30 wt% $NaAlH_4$), MgH_2-50$NaAlH_4$ (50 wt% MgH_2 + 50 wt% $NaAlH_4$), and MgH_2-2Ni-10$NaAlH_4$ (88 wt% MgH_2 + 2 wt% Ni + 10wt% $NaAlH_4$) samples were prepared by milling in a planetary ball mill in hydrogen atmosphere (reactive mechanical milling, RMM). Decomposition temperatures of milled MgH_2, $NaAlH_4$, MgH_2-10$NaAlH_4$, and MgH_2-30$NaAlH_4$ were examined by heating at a rate of 5~6 K/min in a Sieverts-type hydrogen absorption and release apparatus and phase formation in cycled MgH_2-50$NaAlH_4$ were investigated.

2. Experimental Procedures

As starting materials, we used MgH_2 (magnesium hydride, purity 98%, Alfa Aesar, Ward Hill, MA, USA), $NaAlH_4$ (hydrogen storage grade, Aldrich, St. Louis, MO, USA), and Ni (average particle size 2.2–3.0 μm, purity 99.9% metal basis, C typically < 0.1%, Alfa Aesar).

Reactive mechanical milling (RMM) was carried out in a planetary ball mill (Planetary Mono Mill; Pulverisette 6, Fritsch). A mixture with the planned composition (total weight = 8 g) was mixed in a stainless steel container (with 105 hardened steel balls, total weight = 360 g) sealed hermetically, the sample-to-ball weight ratio being 1/45. All samples were handled in a glove box in an argon atmosphere. The disc revolution speed was 400 rpm. The mill container (volume of 250 mL) was then filled with high-purity hydrogen gas of about 12 bar. The RMM was performed for 6 h (by repeating milling for 15 min and pausing for 5 min 24 times). Hydrogen was refilled every 2 h (every eight milling times).

The absorbed or released hydrogen quantity was measured as a function of time by a volumetric method, using the Siverts-type hydrogen absorption and release apparatus previously described [16]. The hydrogen pressure in the reactor was kept at 12 bar during hydriding by dosing the quantity of hydrogen absorbed from a reservoir of known volume. The hydrogen pressure in the reactor was kept at 1.0 bar during dehydriding by removing the quantity of hydrogen released from the reactor to a reservoir of known volume. 0.5 g of the samples was used for these measurements. Samples after hydrogen absorption release cycling were characterized by X-ray diffraction (XRD) with Cu Ka radiation, using a Bruker D8 Advance (Karlsruhe, Germany) powder diffractometer.

3. Results

The quantity of released hydrogen, H_r (wt% H), was calculated using the sample weight as a standard.

Figure 1 shows the quantity of released hydrogen (H_r) versus temperature (T) curves and the ratio of increase in H_r to increase in T, dH_r/dT, versus T curves for milled MgH_2, MgH_2-10$NaAlH_4$, MgH_2-30$NaAlH_4$, and $NaAlH_4$ samples. The samples were heated at a heating rate of 5~6 K/min in 1.0 bar hydrogen. Table 1 presents the temperatures (K) at peaks in the dH_r/dT versus T curves for milled MgH_2, MgH_2-10$NaAlH_4$, MgH_2-30$NaAlH_4$, and $NaAlH_4$ samples. The highest peaks appear at 638, 600, 592, and 455 K, respectively, for milled MgH_2, MgH_2-10$NaAlH_4$, MgH_2-30$NaAlH_4$, and $NaAlH_4$.

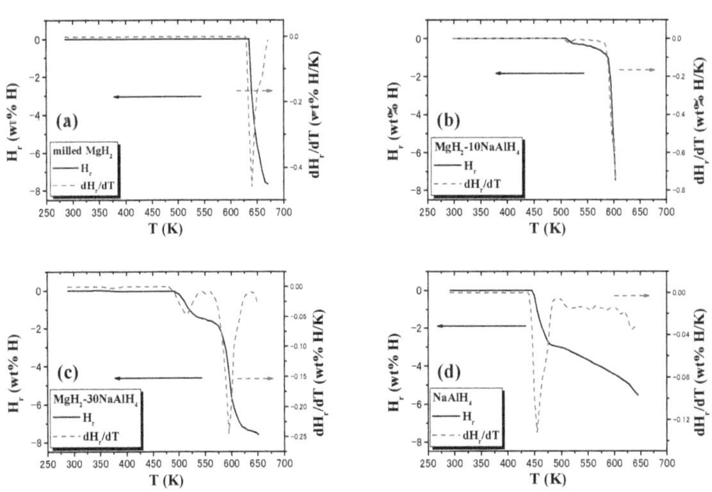

Figure 1. Quantity of released hydrogen (H_r) versus temperature (T) curves and dH_r/dT versus T curves for (**a**) milled MgH_2, (**b**) MgH_2-10$NaAlH_4$, (**c**) MgH_2-30$NaAlH_4$, and (**d**) $NaAlH_4$ samples. The samples were heated at a heating rate of 5~6 K/min in 1.0 bar hydrogen.

Table 1. Temperatures (K) at peaks in the dH_r/dT versus T curves for milled MgH_2, MgH_2-10$NaAlH_4$, MgH_2-30$NaAlH_4$, and $NaAlH_4$ samples.

	Peak	Highest Peak	Peak	Peak
Milled MgH_2		638		
MgH_2-10$NaAlH_4$	513	600		
MgH_2-30$NaAlH_4$	512	592		
$NaAlH_4$		455	552	631

Figure 2 shows the quantity of released hydrogen (H_r) versus temperature (T) curves for milled MgH_2, MgH_2-10$NaAlH_4$, MgH_2-30$NaAlH_4$, and $NaAlH_4$ samples. The samples were heated at a heating rate of 5∼6 K/min in 1.0 bar hydrogen. The points 1, 2, 3, and 4 were marked so that they correspond to the beginning and ending points of the peaks in Figure 1. Table 2 presents the temperatures (K) at the marked points in the H_r versus T curves for milled MgH_2, MgH_2-10$NaAlH_4$, MgH_2-30$NaAlH_4$, and $NaAlH_4$ samples.

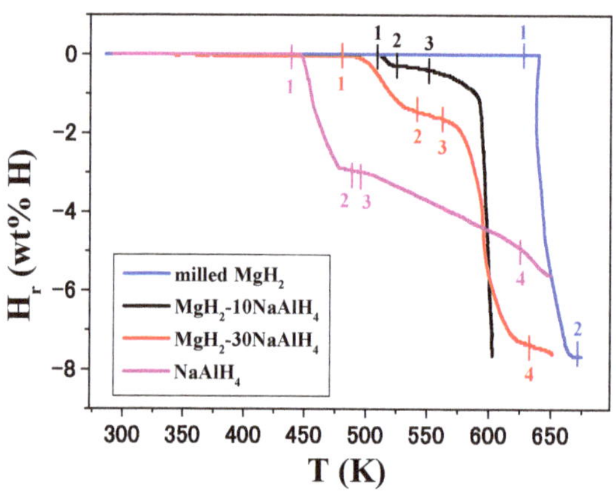

Figure 2. Quantity of released hydrogen (H_r) versus temperature (T) curves for milled MgH_2, MgH_2-10$NaAlH_4$, MgH_2-30$NaAlH_4$, and $NaAlH_4$ samples. The samples were heated at a heating rate of 5∼6 K/min in 1.0 bar hydrogen. The points 1, 2, 3, and 4 were marked so that they correspond to the beginning and ending points of the peaks in Figure 1.

Table 2. Temperatures (K) at the marked points in the H_r versus T curves for milled MgH_2, MgH_2-10$NaAlH_4$, MgH_2-30$NaAlH_4$, and $NaAlH_4$ samples.

Marked Points	1	2	3	4
Milled MgH_2	627	673		
MgH_2-10$NaAlH_4$	508	525	550	
MgH_2-30$NaAlH_4$	480	541	562	633
$NaAlH_4$	438	488	495	625

Hydrogen release begins at 627 K for the as-milled MgH_2. Hydrogen release from the $NaAlH_4$ begins at 438 K and the slope of the H_r versus T curve then changes at 488, 495, and 625 K. MgH_2-10$NaAlH_4$ begins to release hydrogen at 508 K and slopes of the H_r versus T curves change at 525 K and 550 K. MgH_2-30$NaAlH_4$ begins to release hydrogen at 480 K and slopes of the H_r versus T curves change at 541, 562, and 633 K.

As the content of $NaAlH_4$ in the sample increases, the temperature at the highest peak decreases. The higher content of $NaAlH_4$ is believed to strengthen the effects of reactive mechanical grinding, lowering the temperatures for the reaction. The effects of reactive mechanical milling are thought to be the creation of defects, making clean surfaces and cracks, and decreasing particle sizes.

The XRD pattern of MgH_2-50$NaAlH_4$ dehydrided after the number of cycles, n, of 4 at 593 K is shown in Figure 3. The sample contains Al, MgO, Al_3Mg_2, NaH, and $Mg_{17}Al_{12}$. Al, formed from the reaction (4), is believed to react with Mg (formed by the decomposition of MgH_2) and form Al_3Mg_2 and $Mg_{17}Al_{12}$. The $Mg_{17}Al_{12}$ phase belongs to the body-centered cubic system with a space group of I-43m (No. 217). Samples were easily ignited on exposure to the oxygen in the air to obtain the XRD pattern. This led to the formation of a strong peak of MgO and relatively weak peaks of other phases. The broad peak of MgO shows that the particles of MgO are very fine. The large width of the peaks also suggests that the sample has many defects. In the XRD pattern of pure MgH_2, MgO was not found. However, in the XRD pattern of MgH_2-50$NaAlH_4$ after cycling four times, MgO was observed. It is thought that MgO is formed when the sample is exposed to air to obtain the XRD pattern because the particles of the sample are very fine and the surface of the particles are very clean and reactive.

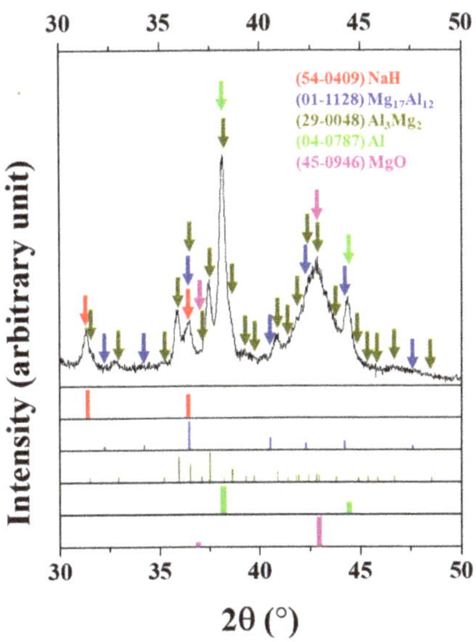

Figure 3. XRD pattern of MgH_2-50$NaAlH_4$ dehydrided after number of cycles, n, of 4 at 593 K. Arrows and bars have the same color as that of a phase.

Liu et al. [22] reported that when the temperature is increased to 633 K, a large amount of hydrogen is released, and two new phases, $Mg_{17}Al_{12}$ and Mg, are formed while the preformed Al and Al_3Mg_2 disappear.

Figure 4 shows the change of H_a versus time t curve at 593 K in 12 bar hydrogen with cycle number, n, for MgH_2-30$NaAlH_4$. The effective hydrogen storage capacity is defined as the quantity of hydrogen absorbed for 60 min (wt% H). As n increases from one to two, the hydrogen absorption rate for 1 min increases. From n = 2 to n = 4, the hydrogen absorption rate for 1 min decreases. In a similar way, the effective hydrogen storage capacity increases as n increases from one to two, and from n = 2 to n = 4, the effective hydrogen storage capacity decreases. The activation is considered to have been completed after n = 2. At n = 2, MgH_2-30$NaAlH_4$ absorbs 4.09 wt% H for 1 min, 7.17 wt% H for 10 min, and 7.42 wt% H for 60 min.

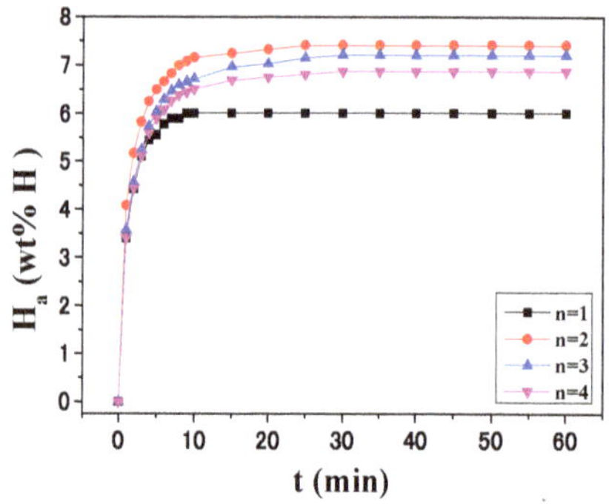

Figure 4. Change in quantity of absorbed hydrogen (H_a) versus time t curve at 593 K in 12 bar hydrogen with cycle number, n, for MgH_2-30$NaAlH_4$.

The change of the H_r versus t curve at 593 K in 1.0 bar hydrogen with n for MgH_2-30$NaAlH_4$ is shown in Figure 5. As the number of cycles (n) increases from one to four, the hydrogen release rate for 1 min decreases. The hydrogen release rate for 1 min at n = 1 and n = 2 are very similar. The quantity of hydrogen released for 60 min increases as n increases from one to two, and from n = 2 to n = 4 the quantity of hydrogen released for 60 min decreases. At n = 2, MgH_2-30$NaAlH_4$ releases 1.31 wt% H for 1 min, 4.95 wt% H for 10 min, and 7.34 wt% H for 60 min.

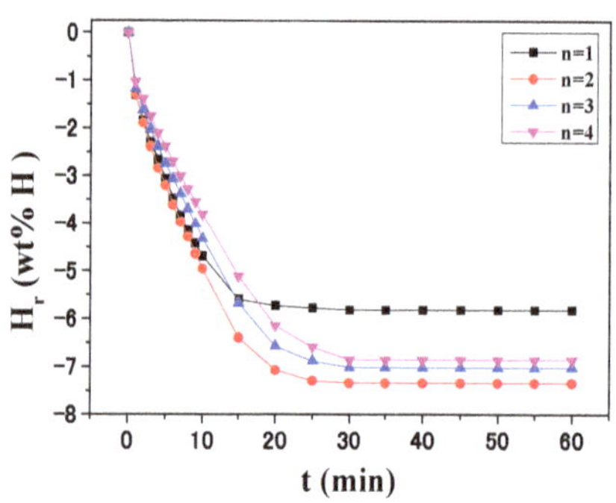

Figure 5. Change in quantity of released hydrogen (H_r) versus t curve at 593 K in 1.0 bar hydrogen with cycle number, n, for MgH_2-30$NaAlH_4$.

Figure 6 shows changes in the H_a versus time t curve at 593 K in 12 bar hydrogen and H_r versus t curve at 593 K in 1.0 bar hydrogen with cycle number, n, for MgH_2-30$NaAlH_4$. The curves show that activation is completed after n = 2, showing the highest hydrogen absorption rate, the highest hydrogen release rate, the largest effective hydrogen storage capacity, and the largest quantity of hydrogen released for 60 min. The curves after completion of activation show that MgH_2-30$NaAlH_4$ has good reversibility and round-trip energy efficiency.

The H_a versus t curves in 12 bar hydrogen at 593 K or 623 K at n = 1 and n = 2 for MgH_2-50$NaAlH_4$ are shown in Figure 7. Because the hydrogen absorption rate was low, experiments were performed several times. Even though the samples were handled in an Ar atmosphere, the samples were ignited partly, leading to low hydrogen absorption rates and low effective hydrogen storage capacities. For n = 1 at 593 K, the MgH_2-50$NaAlH_4$ sample does not absorb hydrogen. For n = 2 at 623 K, the MgH_2-50$NaAlH_4$ sample does not absorb hydrogen either, probably because the difference between the applied hydrogen pressure (12 bar) and the equilibrium plateau pressure at 623 K (6.38 bar [23]) of the Mg-H system is small. At n = 2, the MgH_2-50$NaAlH_4$ sample absorbs 0.99 wt% H for 1 min, 1.36 wt% H for 10 min, and 3.19 wt% H for 60 min at 593 K.

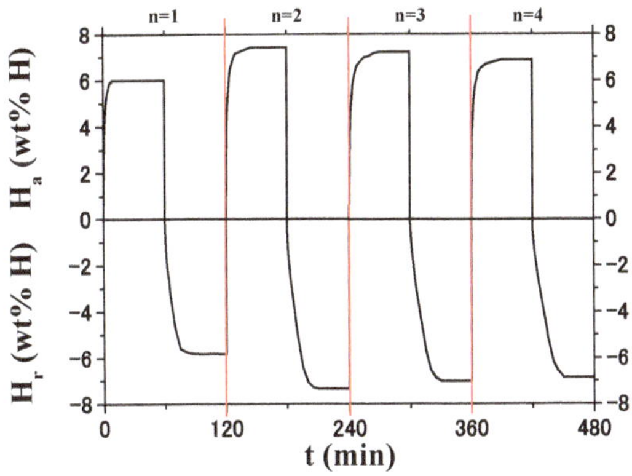

Figure 6. Changes in H_a versus t curve at 593 K in 12 bar hydrogen and H_r versus t curve at 593 K in 1.0 bar hydrogen with cycle number, n, for MgH_2-30$NaAlH_4$.

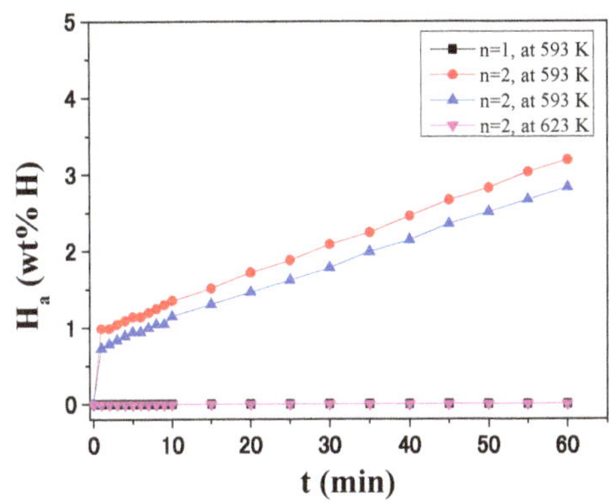

Figure 7. H_a versus t curves in 12 bar hydrogen at 593 K or 623 K at n = 1 and n = 2 for MgH_2-50$NaAlH_4$.

Figure 8 shows the H_r versus t curves in 1.0 bar hydrogen at 593 K or 623 K at n = 1~4 for MgH_2-50$NaAlH_4$. Hydrogen release rates are low and the quantities of hydrogen released for 60 min are small. As n increases, the initial hydrogen release rate and the quantity of hydrogen released for 60 min increase slightly. At n = 2, the MgH_2-50$NaAlH_4$ sample releases 1.03 wt% H for 5 min and 1.29 wt% H for 60 min at 593 K. When the temperature increases from 593 K to 623 K, the initial hydrogen release rate and the quantity of hydrogen released for 60 min increase slightly. Partial ignition in the samples led to low initial hydrogen release rates and small quantities of hydrogen released for 60 min.

During hydrogen absorption (Figure 3) and release (Figure 4), reactions (5) and (6) are believed to occur.

H_a versus t curves in 12 bar hydrogen at 593 K for activated MgH_2-30$NaAlH_4$, MgH_2-2Ni-10$NaAlH_4$ after RMM, activated Mg-10Fe_2O_3 [24,25], activated Mg-10TaF_5 [26,27], and activated Mg-10VCl_3 [28] are shown in Figure 9. The Mg-10Fe_2O_3 [24,25], Mg-10TaF_5 [26,27], and Mg-10VCl_3 [28] samples were also prepared by RMM under conditions similar to those for preparing MgH_2-30$NaAlH_4$ and MgH_2-2Ni-10$NaAlH_4$. MgH_2-2Ni-10$NaAlH_4$ did not require activation after reactive mechanical milling (RMM). The H_a versus t curve of MgH_2-

2Ni-10NaAlH$_4$ after RMM is used for comparison with the H$_a$ versus t curves of other activated samples. Activated MgH$_2$-30NaAlH$_4$ has the highest hydrogen absorption rate for 2.5 min, followed in order by MgH$_2$-2Ni-10NaAlH$_4$ after RMM, activated Mg-10VCl$_3$, activated Mg-10TaF$_5$, and activated Mg-10Fe$_2$O$_3$. Activated MgH$_2$-30NaAlH$_4$ has the highest effective hydrogen storage capacity, followed in order by activated Mg-10VCl$_3$, MgH$_2$-2Ni-10NaAlH$_4$ after RMM, activated Mg-10Fe$_2$O$_3$, and activated Mg-10TaF$_5$. MgH$_2$-30NaAlH$_4$ has a much higher hydrogen absorption rate for 2.5 min (2.20 wt% H/min) and a much larger effective hydrogen storage capacity (7.42 wt% H) than the other samples.

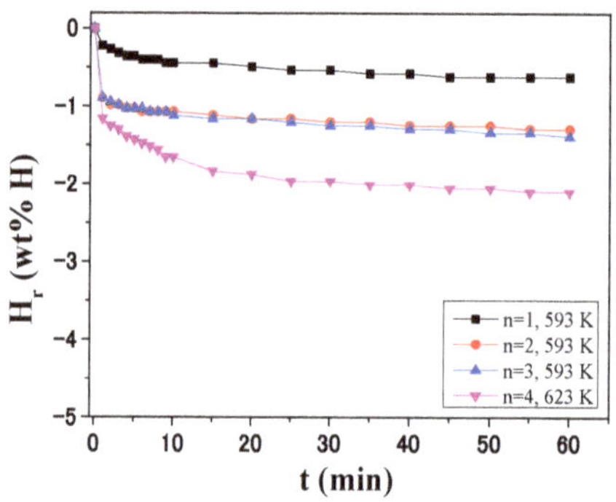

Figure 8. H$_r$ versus t curves in 1.0 bar hydrogen at 593 K or 623 K at n = 1~4 for MgH$_2$-50NaAlH$_4$.

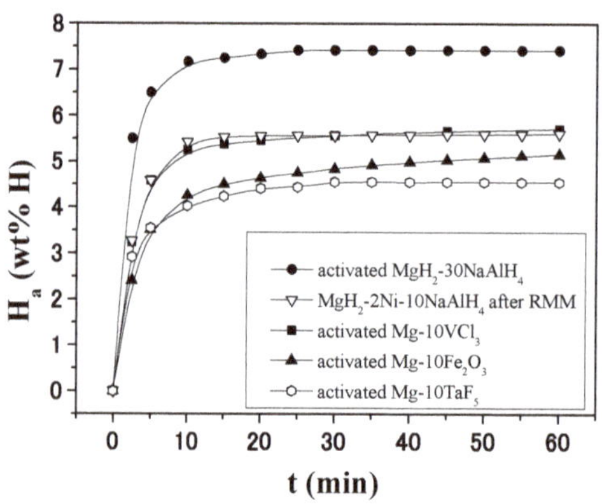

Figure 9. H$_a$ versus t curves in 12 bar hydrogen at 593 K for activated MgH$_2$-30NaAlH$_4$, MgH$_2$-2Ni-10NaAlH$_4$ after RMM, activated Mg-10Fe$_2$O$_3$, activated Mg-10TaF$_5$, and activated Mg-10VCl$_3$.

Figure 10 shows H$_r$ versus t curves in 1.0 bar hydrogen at 593 K for activated MgH$_2$-30NaAlH$_4$, MgH$_2$-2Ni-10NaAlH$_4$ after RMM, activated Mg-10Fe$_2$O$_3$ [24,25], activated Mg-10TaF$_5$ [26,27], and activated Mg-10VCl$_3$ [28]. Activated MgH$_2$-30NaAlH$_4$ has the highest hydrogen release rate for 2.5 min, followed in order by activated Mg-10VCl$_3$, activated Mg-10TaF$_5$, MgH$_2$-2Ni-10NaAlH$_4$ after RMM, and activated Mg-10Fe$_2$O$_3$. Activated MgH$_2$-30NaAlH$_4$ has the largest quantity of hydrogen released for 60 min, followed in order by activated Mg-10VCl$_3$, MgH$_2$-2Ni-10NaAlH$_4$ after RMM, activated Mg-10TaF$_5$, and activated Mg-10Fe$_2$O$_3$, and activated. MgH$_2$-30NaAlH$_4$ has a much higher hydrogen release rate for 2.5 min (0.86 wt%/min) and a much larger quantity of hydrogen released for 60 min (7.34 wt% H) than the other samples.

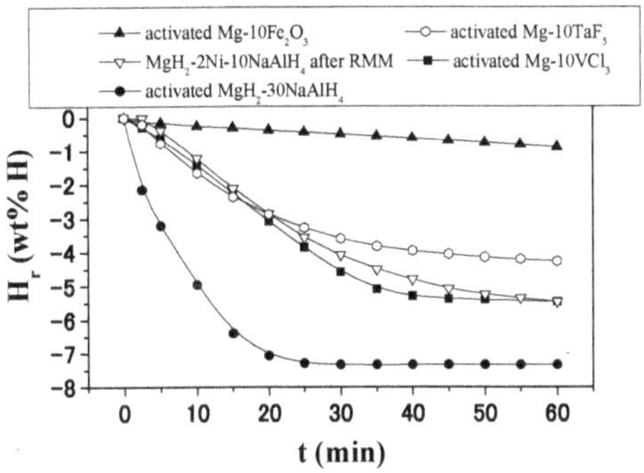

Figure 10. H_r versus t curves in 1.0 bar hydrogen at 593 K for activated MgH_2-30$NaAlH_4$, MgH_2-2Ni-10$NaAlH_4$ after RMM, activated Mg-10Fe_2O_3, activated Mg-10TaF_5, and activated Mg-10VCl_3.

Table 3 shows hydrogen absorption properties in 12 bar hydrogen at 593 K for activated MgH_2-30$NaAlH_4$, MgH_2-2Ni-10$NaAlH_4$ after RMM, activated Mg-10Fe_2O_3 [24], activated Mg-10TaF_5 [26], and activated Mg-10VCl_3 [28]. Numbers in parentheses indicate rankings. H_a (x min) (wt% H) means the quantity of hydrogen absorbed for x min. The initial hydrogen absorption rate is defined as the hydrogen absorption rate during 2.5 min and calculated by dividing H_a (2.5 min) by 2.5 min. Activated MgH_2-30$NaAlH_4$ has the highest initial hydrogen absorption rate, followed in order by MgH_2-2Ni-10$NaAlH_4$ after RMM, activated Mg-10VCl_3, activated Mg-10TaF_5, and activated Mg-10Fe_2O_3. Activated MgH_2-30$NaAlH_4$, MgH_2-2Ni-10$NaAlH_4$ after RMM, and activated Mg-10VCl_3 have relatively large values of H_a (10 min). Activated MgH_2-30$NaAlH_4$ has the highest effective hydrogen storage capacity, followed in order by MgH_2-2Ni-10$NaAlH_4$ after RMM, activated Mg-10VCl_3, activated Mg-10Fe_2O_3, and activated Mg-10TaF_5.

Table 3. Hydrogen absorption properties in 12 bar hydrogen at 593 K for activated MgH_2-30$NaAlH_4$, MgH_2-2Ni-10$NaAlH_4$ after RMM, activated Mg-10Fe_2O_3 (adapted from Ref. [24]), activated Mg-10TaF_5 (adapted from Ref. [26]), and activated Mg-10VCl_3 (adapted from Ref. [28]). Rankings are indicated in parentheses.

Sample	Initial Hydrogen Absorption Rate (wt% H/mn)	H_a (10 min) (wt% H)	Effective Hydrogen Storage Capacity (wt% H)
Activated MgH_2-30$NaAlH_4$	2.20 (1)	7.17 (1)	7.42 (1)
MgH_2-2Ni-10$NaAlH_4$ after RMM	1.31 (2)	5.43 (2)	5.60 (2)
Activated Mg-10Fe_2O_3	0.95 (5)	4.28 (4)	5.16 (4)
Activated Mg-10TaF_5	1.17 (4)	3.82 (5)	4.51 (5)
Activated Mg-10VCl_3	1.26 (3)	5.15 (3)	5.56 (3)

Table 4 shows hydrogen release properties in 1.0 bar hydrogen at 593 K for activated MgH_2-30$NaAlH_4$, MgH_2-2Ni-10$NaAlH_4$ after RMM, activated Mg-10Fe_2O_3 [24], activated Mg-10TaF_5 [26], and activated Mg-10VCl_3 [28]. H_r (x min) (wt% H) means the quantity of hydrogen released for x min. The initial hydrogen release rate is defined as the hydrogen release rate during 2.5 min and calculated by dividing H_r (2.5 min) by 2.5 min. Activated MgH_2-30$NaAlH_4$ has the highest initial hydrogen release rate, followed in order by activated Mg-10VCl_3 and activated Mg-10TaF_5. The initial hydrogen release rates of MgH_2-2Ni-10$NaAlH_4$ after RMM and activated Mg-10Fe_2O_3 are zero. Activated MgH_2-30$NaAlH_4$ has a large value of H_r (10 min). Activated Mg-10TaF_5, activated Mg-10VCl_3, and MgH_2 2Ni-10$NaAlH_4$ after RMM have relatively large values of H_r (10 min). Activated MgH_2-30$NaAlH_4$ has the largest H_r (60 min), followed in order by MgH_2-2Ni-10$NaAlH_4$ after RMM, activated Mg-10VCl_3, activated Mg-10TaF_5, and activated Mg-10Fe_2O_3.

Table 4. Hydrogen release properties in 1.0 bar hydrogen at 593 K for activated MgH_2-30$NaAlH_4$, MgH_2-2Ni-10$NaAlH_4$ after RMM, activated Mg-10Fe_2O_3 (adapted from Ref. [24]), activated Mg-10TaF_5 (adapted from Ref. [26]), and activated Mg-10VCl_3 (adapted from Ref. [28]). Rankings are indicated in parentheses.

Sample	Initial Hydrogen Release Rate (wt% H/mn)	H_r (10 min) (wt% H)	H_r (60 min) (wt% H)
Activated MgH_2-30$NaAlH_4$	0.95 (1)	4.95 (1)	7.34 (1)
MgH_2-2Ni-10$NaAlH_4$ after RMM	0 (4)	1.21 (4)	5.46 (2)
Activated Mg-10Fe_2O_3	0 (4)	0.23 (5)	0.83 (5)
Activated Mg-10TaF_5	0.03 (3)	1.56 (2)	4.01 (4)
Activated Mg-10VCl_3	0.08 (2)	1.32 (3)	5.42 (3)

Table 5 summarizes reversibility, cyclability, round-trip energy efficiency, and general practicability for activated MgH_2-30$NaAlH_4$, MgH_2-2Ni-10$NaAlH_4$ after RMM, activated Mg-10Fe_2O_3, activated Mg-10TaF_5, and activated Mg-10VCl_3. Reversibility (%) is defined as 100 × H_r (60 min)/H_a (60 min) after activation. Cyclability (wt% H/cycle) is defined as the increase in H_a (60 min) per cycle from n =1 to n = 4. Round-trip energy efficiency is defined as H_r (60 min)/H_a (60 min) after activation. General practicability is indicated in ranking by considering all the properties in Tables 3–5. Activated MgH_2-30$NaAlH_4$ has the best reversibility, cyclability, and round-trip energy efficiency, followed in order by MgH_2-2Ni-10$NaAlH_4$ after RMM, activated Mg-10VCl_3, activated Mg-10TaF_5, and activated Mg-10Fe_2O_3. The cyclability of activated MgH_2-30$NaAlH_4$ is positive, showing that H_a (60 min) increases roughly from n = 1 to n = 4.

Table 5. Reversibility, cyclability, round-trip energy efficiency, and general practicability for activated MgH_2-30$NaAlH_4$, MgH_2-2Ni-10$NaAlH_4$ after RMM, activated Mg-10Fe_2O_3, activated Mg-10TaF_5, and activated Mg-10VCl_3. Rankings are indicated in parentheses.

Sample	Reversibility (%)	Cyclability (wt% H/Cycle)	Round-Trip Energy Efficiency	Ranking in General Practicability
Activated MgH_2-30$NaAlH_4$	99 (1)	0.29 (1)	0.99 (1)	1
MgH_2-2Ni-10$NaAlH_4$ after RMM	98 (2)	−0.04 (2)	0.98 (2)	2
Activated Mg-10Fe_2O_3	16 (5)	−0.117 (5)	0.16 (5)	5
Activated Mg-10TaF_5	89 (4)	−0.075 (4)	0.89 (4)	4
Activated Mg-10VCl_3	97 (3)	−0.065 (3)	0.97 (3)	3

Activated MgH_2-30$NaAlH_4$ has the best general practicability, followed in order by MgH_2-2Ni-10$NaAlH_4$ after RMM, activated Mg-10VCl_3, activated Mg-10TaF_5, and activated Mg-10Fe_2O_3. MgH_2-2Ni-10$NaAlH_4$ after RMM has a slightly better general practicability than activated Mg-10VCl_3.

Activated MgH_2-30$NaAlH_4$ has an initial hydrogen absorption rate of 2.20 wt% H/min (vs. 1.26 wt% H/min for activated Mg-10VCl_3), H_a (10 min) of 7.17 wt% H (vs. 5.15 wt% H for activated Mg-10VCl_3), an effective hydrogen storage capacity of 7.42 wt% H (vs. 5.56 wt% H for activated Mg-10VCl_3).

Activated MgH_2-30$NaAlH_4$ has an initial hydrogen release rate of 0.95 wt% H/min (vs. 0.08 wt% H/min for activated Mg-10VCl_3), H_r (10 min) of 4.95 wt% H (vs. 1.32 wt% H for activated Mg-10VCl_3), an H_r (60 min) of 7.34 wt% H (vs. 5.42 wt% H for activated Mg-10VCl_3).

Cyclability in Table 4 is examined only from n = 1 to n = 4. The cyclability of a sample MgH_2-2Ni-10$NaAlH_4$ was examined from n = 1 to n = 30 under 12 bar for hydrogen absorption and 1.0 bar H_2 for hydrogen release at 593 K. MgH_2-2Ni-10$NaAlH_4$ had H_a (60 min) values of 5.62 wt% H at n =1 and 5.36 wt% H at n = 30, showing the good cyclability of −0.009 wt% H/cycle. The lifetimes of the samples were not investigated. We will examine the lifetimes of the samples in future research.

4. Discussion

From the results of Figure 2, it is believed that for the $NaAlH_4$, reaction (1) (decomposition of $NaAlH_4$) begins to occur at 438 K, and reaction (2) (decomposition of Na_3AlH_6) begins to occur at 495 K. For the MgH_2-10$NaAlH_4$ sample, it is believed that reaction (4) occurs between point 1 (508 K) and point 2 (525 K), reactions (5) and (6) begin to occur after point 3 (550 K), and then reaction (7) and reaction (8) occur consecutively. For the MgH_2-30$NaAlH_4$ sample, it is believed that reaction (4) occurs between point 1 (480 K) and point 2 (541 K), reactions (5) and (6) begin to occur after point 3 (562 K), and then reaction (7) and reaction (8) occur consecutively. The reaction (4) for the MgH_2-30$NaAlH_4$ begins to occur at 28 K lower temperature than that for the MgH_2-10$NaAlH_4$.

Samples were easily ignited on exposure to air, making the obtention of XRD patterns difficult and leading to the formation of a strong peak of MgO and relatively weak peaks of other phases. Liu et al. [22] reported that when the temperature is increased to 633 K, the preformed Al and Al_3Mg_2 disappear. In Figure 3 (the XRD pattern of MgH_2-50$NaAlH_4$ dehydrided after the number of cycles, n, of 4 at 593 K), the Al_3Mg_2 phase is observed. The dehydriding temperature (593 K) lower than that in the work of Liu et al. [22] is believed to have led to this result.

As the content of $NaAlH_4$ in the sample increases, the temperature at the highest peak in the ratio of increase in H_r to increase in T, dH_r/dT, versus T curve decreases. The higher content of $NaAlH_4$ is believed to strengthen the effects of reactive mechanical milling. However, too much content of $NaAlH_4$ (as in MgH_2-50$NaAlH_4$) leads to worse hydrogen storage properties (hydrogen absorption rate, hydrogen release rate, and hydrogen storage capacity).

From the results of this work and the reported study [20], it is believed that hydriding in 12 bar hydrogen and dehydriding in 1.0 bar hydrogen at 593 K of MgH_2-30$NaAlH_4$ are performed by the reversible reactions $MgH_2 \Leftrightarrow Mg + H_2$ and $17MgH_2 + 12Al \Leftrightarrow Mg_{17}Al_{12} + 17H_2$.

In a planetary ball mill, each jar is located on a circular platform (sun wheel). When the sun wheel turns, the jar rotates around its own axis in the opposite direction. The resulting centrifugal and acting acceleration forces lead to strong grinding effects. RMM can break thin surface oxides and expose fresh metallic surfaces, which are reactive. RMM can also introduce defects and cracks. The propagation of cracks, due to additives acting as the center of stress, makes the particles finer. Defects can be used as the sites active for nucleation. The exposed fresh metallic surfaces and created cracks and surfaces have high reactivity. A decrease in particle size reduces the diffusion distance of hydrogen atoms. The added materials and formed phases are believed to strengthen the effects of RMM. The expansion of the lattice due to hydrogen absorption and the contraction of the lattice due to hydrogen release causes effects similar to those of RMM. However, the effects of lattice expansion and contraction will be weaker than those of RMM. The MgH_2-30$NaAlH_4$ sample has a higher hydrogen absorption rate for 2.5 min, a larger effective hydrogen storage capacity, a higher hydrogen release rate for 2.5 min, and a larger quantity of hydrogen released for 60 min than the other samples, showing that the effects of RMM and hydrogen absorption release cycling are stronger in the MgH_2-30$NaAlH_4$ sample, compared with those for the other samples. Reportedly, nucleation can be facilitated by creating active nucleation sites by mechanical treatment and/or alloying with additives [29]; the diffusion distance of hydrogen can also be decreased by the mechanical treatment and/or alloying of Mg with additives, thereby reducing the magnesium particle size [30]. In addition, hydrogen mobility can be improved by additives that create microscopic paths of hydrogen [30]; a rough surface of magnesium particles having many cracks and defects is thus considered more advantageous for hydrogen absorption [31].

In our future work, milled $NaAlH_4$ will be prepared by reactive mechanical milling. The quantity of released hydrogen (H_r) versus temperature T curve and the ratio of increase in H_r to increase in T, dH_r/dT, versus the T curve for the milled $NaAlH_4$, will be obtained and studied in detail. Behaviors of the milled $NaAlH_4$, which are different from those of $NaAlH_4$, are expected.

As shown in Figures 9 and 10, the additional effects of $NaAlH_4$, oxide, halides, or fluoride on MgH_2 or Mg are different. Which kinds of properties such as physical properties (hardness, toughness, surface area, microstructure, etc.) and chemical properties affect the hydrogen storage properties will be investigated in future work.

5. Conclusions

In the present work, milled MgH_2, MgH_2-10$NaAlH_4$ (with a composition of 90 wt% MgH_2 + 10 wt% $NaAlH_4$), MgH_2-30$NaAlH_4$ (70 wt% MgH_2 + 30 wt% $NaAlH_4$), MgH_2-50$NaAlH_4$ (50 wt% MgH_2 + 50 wt% $NaAlH_4$), and MgH_2-2Ni-10$NaAlH_4$ (88 wt% MgH_2 + 2 wt% Ni + 10wt% $NaAlH_4$) samples were prepared by reactive mechanical milling (RMM). Effects of the $NaAlH_4$ content on the temperatures of intermediate reactions were studied for MgH_2-$NaAlH_4$ composites. Phase formation in the cycled MgH_2-50$NaAlH_4$ was investigated. The hydrogen storage properties of MgH_2-30$NaAlH_4$ were compared with those of Mg-based alloys in which oxide, halides, or fluoride were added. Hydrogen release begins at 627 K for the as-milled MgH_2. MgH_2-30$NaAlH_4$ begins to release hydrogen at 480 K and slopes of the H_r versus T curves change at 541, 562, and 633 K. As the content of $NaAlH_4$ in the sample increased, the temperature at the highest peak in the ratio of increase in H_r to increase in T, dH_r/dT, versus T curve decreased. The higher content of $NaAlH_4$ is believed to have strengthened the effects of reactive mechanical milling. MgH_2-50$NaAlH_4$ dehydrided after four cycles contained Al, MgO, Al_3Mg_2, NaH, and $Mg_{17}Al_{12}$. Hydriding in 12 bar hydrogen and dehydriding in 1.0 bar hydrogen at 593 K of MgH_2-30$NaAlH_4$ are performed by the reversible reactions $MgH_2 \Leftrightarrow Mg + H_2$ and $17MgH_2 + 12Al \Leftrightarrow Mg_{17}Al_{12} + 17H_2$. Activation of MgH_2–30$NaAlH_4$ was completed after two hydrogen absorption release cycles. MgH_2–30$NaAlH_4$ was the best Mg-based composite among Mg-based alloys in which an oxide, a halide, a fluoride, or a complex hydride was added, with a high hydrogen absorption rate for 2.5 min (2.20 wt% H/min) and a large effective hydrogen storage capacity (7.42 wt% H).

Author Contributions: Conceptualization, Y.-J.K., M.-Y.S. and K.-T.L.; methodology, Y.-J.K., M.-Y.S. and K.-T.L.; formal analysis, Y.-J.K., M.-Y.S. and K.-T.L.; investigation, Y.-J.K., M.-Y.S. and K.-T.L.; data curation, Y.-J.K.; writing—original draft preparation, Y.-J.K. and M.-Y.S.; writing—review and editing—Y.-J.K., M.-Y.S. and K.-T.L.; project administration, Y.-J.K. and K.-T.L.; funding acquisition, Y.-J.K. All authors have read and agreed to the published version of the manuscript.

Funding: This work was supported by the National Research Foundation of Korea (NRF) grant funded by the Korean government (MSIT) (No. 2021R1C1C2009103). This work was also supported by Korea Institute of Energy Technology Evaluation and Planning (KETEP) grant funded by the Korean government (MOTIE) (20213030040110).

Data Availability Statement: The original contributions presented in the study are included in the article, further inquiries can be directed to the corresponding author, M.-Y.S.

Conflicts of Interest: The authors declare no conflicts of interest.

References

1. Pratthana, C.; Aguey-Zinsou, K.F. Surfactant Induced Synthesis of $LiAlH_4$ and $NaAlH_4$ Nanoparticles for Hydrogen Storage. *Appl. Sci.* **2022**, *12*, 4742. [CrossRef]
2. Chen, X.; Li, Z.; Zhang, Y.; Liu, D.; Wang, C.; Li, Y.; Si, T.; Zhang, Q. Enhanced Low-Temperature Hydrogen Storage in Nanoporous Ni-Based Alloy Supported $LiBH_4$. *Front. Chem.* **2020**, *8*, 283. [CrossRef] [PubMed]
3. Guo, Y.; Liu, Y.; Feng, L.; An, C.; Wang, Y. High-loading $LiBH_4$ Confined in Structurally Tunable Ni Catalyst-decorated Porous Carbon Scaffold for Fast Hydrogen Desorption. *Chem. Asian J.* **2023**, *18*, 1. [CrossRef]
4. Chen, W.; Sun, Y.; Xu, T.; Ye, J.; Xia, G.; Sun, D.; Yu, X. Reversible Hydrogen Storage Performance of $2LiBH_4$-MgH_2 Enabled by Dual Metal Borides. *ACS Appl. Energy Mater.* **2022**, *5*, 10501. [CrossRef]
5. Lin, W.; Xiao, X.; Wang, X.; Wong, J.W.; Yao, Z.; Chen, M.; Zheng, J.; Hu, Z.; Chen, L. Extreme high reversible capacity with over 8.0 wt% and excellent hydrogen storage properties of MgH_2 combined with $LiBH_4$ and Li_3AlH_6. *J. Energy Chem.* **2020**, *50*, 296. [CrossRef]

6. Huang, X.; Xiao, X.; He, Y.; Yao, Z.; Ye, X.; Kou, H.; Chen, C.; Huang, T.; Fan, X.; Chen, L. Probing an intermediate state by X-ray absorption near-edge structure in nickel-doped $2LiBH_4$–MgH_2 reactive hydride composite at moderate temperature. *Mater. Today Nano* **2020**, *12*, 100090. [CrossRef]
7. Shen, S.; Liao, W.; Cao, Z.; Liu, J.; Wang, H.; Ouyang, L. Enhanced hydrogen storage properties of MgH_2 with the co-addition of $LiBH_4$ and YNi_5 alloy. *J. Mater. Sci. Technol.* **2024**, *178*, 90. [CrossRef]
8. Shang, Y.; Pistidda, C.; Milanese, C.; Girella, A.; Schökel, A.; Le, T.T.; Hagenah, A.; Metz, O.; Klassen, T.; Dornheim, M. Sustainable $NaAlH_4$ production from recycled automotive Al alloy. *Green Chem.* **2022**, *24*, 4153. [CrossRef]
9. Li, Z.; Yu, J.Z.; Zhang, Y.; Liu, D.M.; Wang, C.Y.; Si, T.Z.; Li, Y.T.; Zhang, Q.A. Coupling of nanoconfinement with metallic catalysis in supported $NaAlH_4$ for low-temperature hydrogen storage. *J. Power Sources* **2021**, *491*, 229611. [CrossRef]
10. Chen, J.; Li, C.; Chen, W.; Zhang, X.; Yu, X.; Xia, G. Tailoring reversible hydrogen storage performance of $NaAlH_4$ through $NiTiO_3$ nanorods. *J. Alloys Compd.* **2024**, *971*, 172689. [CrossRef]
11. Reilly, J.J.; Wiswall, R.H. Reaction of hydrogen with alloys of magnesium and nickel and the formation of Mg_2NiH_4. *Inorg. Chem.* **1968**, *7*, 2254–2256. [CrossRef]
12. Karty, A.; Genossar, J.G.; Rudman, P.S. Hydriding and dehydriding kinetics of Mg in a Mg/Mg_2Cu eutectic alloy: Pressure sweep method. *J. Appl. Phys.* **1979**, *50*, 7200–7209. [CrossRef]
13. Akiba, E.; Nomura, K.; Ono, S.; Suda, S. Kinetics of the reaction between Mg-Ni alloys and H_2. *Int. J. Hydrogen Energy* **1982**, *7*, 787–791. [CrossRef]
14. Bobet, J.-L.; Akiba, E.; Nakamura, Y.; Darriet, B. Study of Mg–M (M = Co, Ni and Fe) mixture elaborated by reactive mechani cal alloying-hydrogen sorption properties. *Int. J. Hydrogen Energy* **2000**, *25*, 987–996. [CrossRef]
15. Huot, J.; Ravnsbæk, D.B.; Zhang, J.; Cuevas, F.; Latroche, M.; Jensen, T.R. Mechanochemical synthesis of hydrogen storage materials. *Prog. Mater. Sci.* **2013**, *58*, 30–75. [CrossRef]
16. Song, M.Y.; Ahn, D.S.; Kwon, I.H.; Ahn, H.J. Development of Hydrogen storage Alloys by Mechanical Alloying Mg with Fe and Co. *Met. Mater. Int.* **1999**, *5*, 485–490. [CrossRef]
17. Ali, N.A.; Ismail, M. Advanced hydrogen storage of the Mg–Na–Al system: A review. *J. Magnes. Alloys* **2021**, *9*, 1111–1122. [CrossRef]
18. Plerdsranoy, P.; Meethom, S.; Utke, R. Dehydrogenation kinetics, reversibility, and reaction mechanisms of reversible hydrogen storage material based on nanoconfined $MgH_2-NaAlH_4$. *J. Phys. Chem. Solids* **2015**, *87*, 16–22. [CrossRef]
19. Rafi-ud-din; Xuanhui, Q.; Ping, L.; Zhang, L.; Ahmad, M.; Iqbal, M.Z.; Rafique, M.Y.; Farooq, M.H. Enhanced hydrogen storage performance for MgH_2–$NaAlH_4$ system—The effects of stoichiometry and Nb_2O_5 nanoparticles on cycling behaviour. *RSC Adv.* **2012**, *2*, 4891–4903. [CrossRef]
20. Ismail, M.; Zhao, Y.; Yu, X.B.; Mao, J.F.; Dou, S.X. The hydrogen storage properties and reaction mechanism of the MgH_2–$NaAlH_4$ composite system. *Int. J. Hydrogen Energy* **2011**, *36*, 9045–9050. [CrossRef]
21. Bendyna, J.K.; Dyjak, S.; Notten, P.H.L. The influence of ball-milling time on the dehydrogenation properties of the $NaAlH_4$–MgH_2 composite. *Int. J. Hydrogen Energy* **2015**, *40*, 4200–4206. [CrossRef]
22. Liu, H.; Wang, X.; Liu, Y.; Dong, Z.; Ge, H.; Li, S.; Yan, M. Hydrogen Desorption Properties of the MgH_2–AlH_3 Composites. *J. Phys. Chem. C* **2014**, *118*, 37–45. [CrossRef]
23. Stampfer, J.F.; Molley, C.E.; Suttle, J.F. The Magnesium-Hydrogen System1-3. *J. Am. Chem. Soc.* **1960**, *82*, 3504–3508. [CrossRef]
24. Song, M.Y.; Kwon, I.H.; Kwon, S.N.; Park, C.G.; Hong, S.-H.; Bae, J.-S.; Mumm, D.R. Hydrogen-storage properties of Mg–oxide alloys prepared by reactive mechanical grinding. *J. Alloys Compd.* **2006**, *415*, 266–270. [CrossRef]
25. Song, M.Y.; Kwon, S.N.; Park, H.R. Pressure-Composition Isotherms and Cycling Properties of Mg-xFe_2O_3-yNi Alloys. *Korean J. Met. Mater.* **2013**, *51*, 455–460. [CrossRef]
26. Kwak, Y.J.; Lee, S.H.; Park, H.R.; Song, M.Y. Review Paper: Hydriding and Dehydriding Reactions of Mg-xTaF_5 (x=0, 5, and 10) Prepared via Reactive Mechanical Grinding. *Korean J. Met. Mater.* **2014**, *52*, 957–962.
27. Kwak, Y.J.; Lee, S.H.; Park, H.R.; Song, M.Y. Hydrogen Storage Characteristics of Mg, Mg-5TaF_5, and Mg-5NbF_5 Prepared via Grinding in a Hydrogen Atmosphere. *J. Nanosci. Nanotechnol.* **2016**, *16*, 10508–10514. [CrossRef]
28. Song, M.Y.; Lee, S.H.; Mumm, D.R. Fe_2O_3, MnO, and VCl_3-added Mg composites by reaction involving grinding processing for hydrogen storage. *J. Ceram. Proc. Res.* **2018**, *19*, 211–217.
29. Hjort, P.; Krozer, A.; Kasemo, B. Hydrogen sorption kinetics in partly oxidized Mg films. *J. Alloys Comp.* **1996**, *237*, 74–80. [CrossRef]
30. Zaluska, A.; Zaluski, L.; Ström-Olsen, J.O. Nanocrystalline magnesium for hydrogen storage. *J. Alloys Comp.* **1999**, *288*, 217–225. [CrossRef]
31. Vigeholm, B.; Kjoller, J.; Larsen, B.; Pedersen, A.S. Formation and decomposition of magnesium hydride. *J. Less-Common Met.* **1983**, *89*, 135–144. [CrossRef]

metals

MDPI

Article

Inverse Magnetocaloric Effect in Heusler $Ni_{44.4}Mn_{36.2}Sn_{14.9}Cu_{4.5}$ Alloy at Low Temperatures

Alexander P. Kamantsev [1,*], Yuriy S. Koshkid'ko [2], Ruslan Yu. Gaifullin [3], Irek I. Musabirov [3], Anatoliy V. Koshelev [4], Alexey V. Mashirov [1], Vladimir V. Sokolovskiy [5,6], Vasiliy D. Buchelnikov [5], Jacek Ćwik [2] and Vladimir G. Shavrov [1]

1 Kotelnikov Institute of Radioengineering and Electronics of Russian Academy of Sciences, 125009 Moscow, Russia; mashirov@cplire.ru (A.V.M.); shavrov@cplire.ru (V.G.S.)
2 Institute of Low Temperature and Structure Research of Polish Academy of Sciences, 50-422 Wroclaw, Poland; y.koshkidko@intibs.pl (Y.S.K.); j.cwik@intibs.pl (J.Ć.)
3 Institute for Metals Superplasticity Problems of Russian Academy of Sciences, 450001 Ufa, Russia; gaifullin_1998@bk.ru (R.Y.G.); irekmusabirov@imsp.ru (I.I.M.)
4 Institute of Experimental Mineralogy of Russian Academy of Sciences, 142432 Chernogolovka, Russia; anatolkosh@mail.ru
5 Department of Condensed Matter Physics, Chelyabinsk State University, 454001 Chelyabinsk, Russia; vsokolovsky84@mail.ru (V.V.S.); buche@csu.ru (V.D.B.)
6 Academic Research Center for Energy Efficiency, National University of Science and Technology "MISIS", 119049 Moscow, Russia
* Correspondence: kama@cplire.ru

Abstract: Direct measurements of the magnetocaloric effect were performed in a Heusler $Ni_{44.4}Mn_{36.2}Sn_{14.9}Cu_{4.5}$ alloy at cryogenic temperatures in magnetic fields up to 10 T. The maximum value of the inverse magnetocaloric effect in a 10 T field was ΔT_{ad} = −2.7 K in the vicinity of the first-order magnetostructural phase transition at T_0 = 117 K. Ab initio and Monte Carlo calculations were performed to discuss the effect of Cu doping into a Ni-Mn-Sn compound on the ground-state structural and magnetic properties. It is shown that with increasing Cu content the martensitic transition temperature decreases and the Curie temperature of austenite slightly increases. In general, the calculated transition temperatures and magnetization values correlated well with the experimental ones.

Keywords: metamagnetic phase transition; magnetocaloric effect; Heusler alloys; Monte Carlo calculations

1. Introduction

The hopeful idea of solid-state magnetic cooling (SMC) through the adiabatic demagnetization of paramagnetic Gd salts was used in 1933 to develop novel methods to achieve low temperatures below 1 K [1]. SMC is based on the magnetocaloric effect (MCE), which is a reversible change in the temperature ΔT_{ad} (under adiabatic conditions) or entropy ΔS_{iso} (under isothermal conditions) of a magnetic material due to external magnetic field changes [2,3]. The MCE is an efficient apparatus for the investigation of magnetic phase transitions (PTs), which are among the most exigent problems in modern solid-state physics. The maximum MCE is reached in the vicinity of magnetic PTs [3,4]. Magnetic materials with magnetic PTs in the required working temperature range are chosen for SMC. The promising applications of SMC at low temperatures, particularly for the liquefaction of gases such as N_2, He, H_2 or natural gases, have been mentioned in [5–8]. The advantage of such materials for SMC at low temperatures is connected with the increase in their magnetic heat capacity in the magnetic PT region. It becomes comparable to (and sometimes exceeds) the crystal lattice heat capacity. This fact makes SMC more promising at low temperatures, where the lattice heat capacity of metals is much higher than that at room temperature. It is currently accepted that low (cryogenic) temperatures are those below 120 K [9].

The MCE can be either direct or inverse under the magnetization of a magnetic material [2–4]. The direct MCE is an increase in temperature at adiabatic conditions or a decrease in entropy at isothermal conditions. Conversely, the inverse MCE is a positive isothermal change in entropy or a negative adiabatic change in temperature under magnetization. Both the direct and inverse MCE can be applied for SMC at low temperatures. The inverse MCE can be applicable when it is necessary to quickly cool a magnetic field area. For this purpose, a magnetic field is simply applied to the working body. Thus, the working body will cool down, and it will also cool the magnetic field area. This effect can be applied to the cooling of a superconducting magnet and the surrounding liquefied gas, making it possible to reduce the expenditure of liquefied gas.

The rare-earth metals [4] and their intermetallic compounds [10] are the most promising materials for SMC at low temperatures. The magnetostructural PTs in such alloys are commonly observed at temperatures above 120 K [2]. Most often, they show a direct MCE in the vicinity of the PTs. The inverse MCE is observed in their intermetallic compounds with antiferromagnetic (AFM) ordering: RCu_2, R_2Fe_{17}, RFe_3 (R—heavy rare-earth metal) [10–12], Gd-based alloys (Gd_2In [13], GdRuSi [14]). The inverse MCE can also be observed as a result of magnetization rotation upon the magnetization of a highly anisotropic single-crystal sample along the hard magnetization axis in, for example, RCo_5 [15] and Tb_2CoMnO_6 [16]. However, the rare-earth metals are expensive, and their compounds are breakable and rapidly oxidize in air, so they are difficult to use in the proposed SMC scheme.

Significant inverse MCE can be observed in several ferromagnetic (FM) Heusler alloys based on Ni–Mn–Z (Z = In, Sn, Sb), but they usually show the effect closer to room temperature [17–22]. Some Heusler alloy compositions show the inverse MCE at low temperatures, for example, the Ni(-Co)-Mn-Ti Heusler alloys [23]. In general, the Heusler alloys are cheap to produce, as they consist of inexpensive chemical elements and do not require long-term heat treatment, unlike alloys based on expensive rare-earth elements. Therefore, the goal of our work was to study the inverse MCE at low temperatures in a Heusler composition, which was chosen according to [24]. We investigated a $Ni_{44.4}Mn_{36.2}Sn_{14.9}Cu_{4.5}$ Heusler alloy with the inverse MCE at low temperature for SMC and try to explain its properties theoretically in this paper.

2. Materials and Methods

2.1. Samples Characterization

A polycrystalline ingot of the $Ni_{44.4}Mn_{36.2}Sn_{14.9}Cu_{4.5}$ Heusler alloy was synthesized via the argon arc-melting method from high-purity elements. The ingot was remelted seven times to eliminate the chemical segregation of the composition. The obtained ingot had a mass of about 80 g. The samples, in the form of 1 mm thick plates, were placed in quartz ampoules, in which a vacuum of about 10 Pa was created. The ampoules were then sealed. Previous research found that the melting point of this alloy is 1313 K [25]; therefore, the temperature of the homogenization annealing was selected as 1133 K. The sealed samples were subjected to homogenization annealing for 24 h, and then they were quenched in water.

The quenched plates were prepared for structure analysis by electropolishing in an electrolyte with 90% n-butyl alcohol (C_4H_9OH) and 10% HCl. X-ray diffraction and X-ray phase analyses were performed on a "Rigaku" Ultima V X-ray diffractometer using Cu-Kα radiation. The crystal structure of the studied sample at room temperature was cubic (space group *Fm-3m*), with lattice parameter a_0 = 5.991 Å. The analyses of the microstructure and elemental composition were carried out on a "Tescan" Vega 3-SBH scanning electron microscope (SEM) equipped with sensors for backscattered electrons and energy-dispersive X-ray (EDX) analysis on a "Oxford Instruments" X-Act. The specimens were prepared by electropolishing in an electrolyte with 90% n-butyl alcohol (C_4H_9OH) and 10% HCl. The microstructure of the alloy as observed via SEM is represented by a single-phase state in Figure 1a. The dark points on the thin section are etching pits. The grains have a size on the order of 100 μm, according to the orientational contrast. The microstructure can generally be characterized as equiaxed. The unblurred contrast means that the grain boundaries are high-angle. The chemical element maps of the section area obtained by EDX analysis are presented in Figure 1b. The distribution of the Ni, Mn, Sn and Cu is uniform, and there are no areas with the localization of any elements. It can be concluded that annealing at 1133 K for 24 h followed by quenching made it possible to obtain a single-phase-state sample, although this determination is constrained by the limitations of EDX analysis.

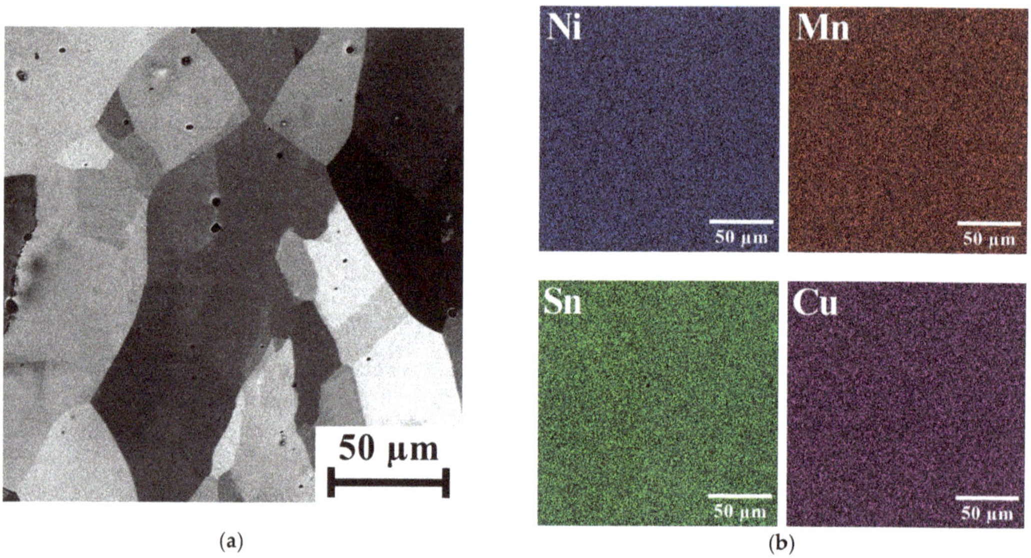

Figure 1. (**a**) Microstructure of the Ni44.4Mn36.2Sn14.9Cu4.5 Heusler alloy, obtained via SEM. (**b**) Maps of the chemical element distribution of the section area of the $Ni_{44.4}Mn_{36.2}Sn_{14.9}Cu_{4.5}$ Heusler alloy, obtained via EDX analysis.

The magnetic properties of the $Ni_{44.4}Mn_{36.2}Sn_{14.9}Cu_{4.5}$ Heusler alloy samples were studied using standard magnetometry methods (ZFC-FC-FH protocols) using a "Quantum Design" MPMS-7T SQUID magnetometer in a low magnetic field of 10 mT (inset in Figure 2a) and in fields of 1 T and 3 T in the wide temperature range of 2–400 K (Figure 2a). The alloy demonstrated two magnetic PTs of first and second order. The 1st-order magnetostructural PT from a ferrimagnetic (FiM) martensitic phase to an FM austenitic phase was observed at the cryogenic temperatures. The characteristic start and finish temperatures of the martensite and austenite states obtained via the tangential method are denoted as M_S = 101 K, M_F = 54 K, A_S = 70 K, and A_F = 118 K, respectively. A temperature hysteresis of about 20 K was observed during the heating/cooling process (inset in Figure 2a). The application of high magnetic fields shifted the region of the martensitic phase to lower temperatures at the rate of −2.2 K/T (Figure 2a), which made it possible to realize the inverse MCE in this alloy in relatively low magnetic fields at cryogenic temperatures. The 2nd order magnetic PT was at the Curie temperature, $T_C \approx 350$ K.

Figure 2b shows the magnetic field dependence of the magnetization of the $Ni_{45.3}Mn_{35.9}Sn_{14.3}Cu_{4.5}$ Heusler alloy at different temperatures in the vicinity of the magnetostructural PT from 50 to 130 K, with increasing and decreasing magnetic field strengths. Measurements of the field dependences of magnetization were carried out using a self-build vibrating magnetometer with a stepper motor [26] in a Bitter magnet field up to 10 T. The sample temperature was established and maintained during each measurement. Before each measurement, the sample was cooled down to 4.2 K without a magnetic field, i.e., below the M_F temperature. This was carried out to remove residual austenite phase that may have formed as a result of the previous magnetization/demagnetization processes.

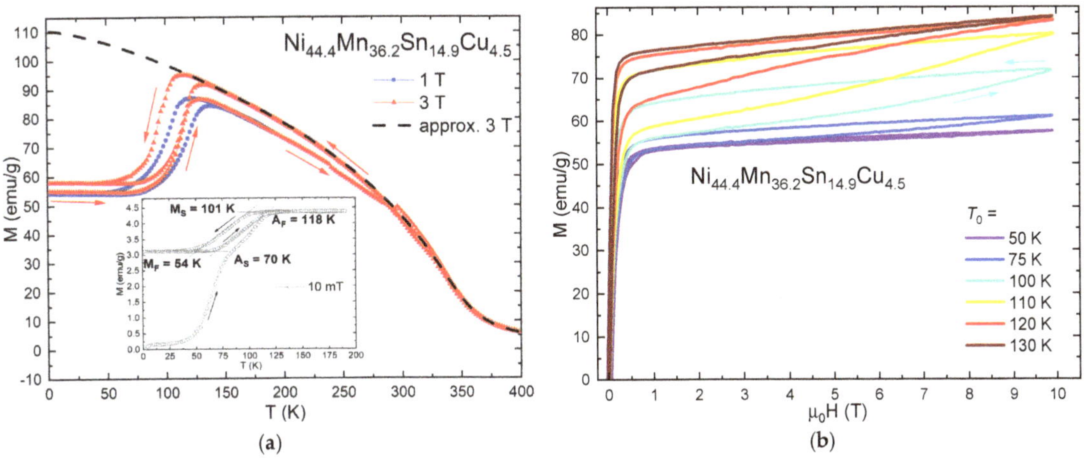

Figure 2. (**a**) Temperature dependences of the magnetization of the $Ni_{45.3}Mn_{35.9}Sn_{14.3}Cu_{4.5}$ Heusler alloy in high magnetic fields of 1 and 3 T (inset: in low field of 10 mT). The dashed line is the calculated curve (approximation) of the temperature dependence of the magnetization in a field of $\mu_0 H = 3$ T for the sample in austenite phase (without a magnetostructural PT). (**b**) Field dependences of the magnetization of the $Ni_{45.3}Mn_{35.9}Sn_{14.3}Cu_{4.5}$ Heusler alloy at different temperatures from 50 to 130 K, obtained with increasing and decreasing field strength.

2.2. Computational Methods

In order to study the role of Cu addition on the structural and magnetic properties of Ni-Mn-Sn alloy, we also carried out theoretical studies within the framework of ab initio calculations and Monte Carlo (MC) simulations.

First-principles calculations of the ground-state properties of the austenitic and martensitic phases of Cu-doped Ni-Mn-Sn alloys were performed using the projector augmented wave (PAW) method implemented in the Vienna Ab initio Simulation Package (VASP ver. 6.2) [27,28]. The generalized gradient approximation in the formulation of Perdew, Burke, and Ernzerhof [29] was chosen to calculate the exchange-correlation energy. Electron–ion interactions are described using PAW potentials with the electronic configurations as follows: $3p^6 3d^8 4s^2$ for Ni, $3p^6 3d^6 4s^1$ for Mn, $3d^{10} 4s^2 4p^2$ for Sn, and $3p^6 3d^{10} 4s^1$ for Cu. The cutoff energy of plane waves was 450 eV. The Brillouin zone was sampled using a $5 \times 5 \times 4$ mesh of k points centered at the Gamma point. The convergence criterion for the total energy was 10^{-6} eV/at. A $2 \times 2 \times 2$ supercell of 64 atoms with an 8-atom tetragonal unit cell ($L1_0$, I4/mmm, #139) was considered.

The geometric optimization procedure was carried out for the compounds $Ni_{32-x}Cu_xMn_{23}Sn_9$ (x = 1, 2, and 3), which correspond to $Ni_{50-x}Cu_xMn_{35.94}Sn_{14.06}$ (x = 1.56, 3.12 and 4.69 at.%), where the latter is close to the experimental $Ni_{44.4}Cu_{4.5}Mn_{36.2}Sn_{14.9}$. To form $Ni_{50-x}Cu_xMn_{35.94}Sn_{14.06}$ (x = 1.56, 3.12 and 4.69 at.%), the Mn_2 and Cu atoms were randomly distributed at Sn and Ni sites. Two magnetic configurations were considered: FM, where the Mn_1, Mn_2 and Ni magnetic moments were aligned in parallel, and FiM, where the magnetic moments of Mn_2 atoms were reversed with respect to the Mn_1 and Ni magnetic moments. Here Mn_1 and Mn_2 are Mn atoms located in the regular Mn sublattice and Sn sublattice, respectively.

The spin-polarized relativistic Korringa–Kohn–Rostoker package (SPR-KKR ver. 7.7) implemented in [30] was applied for calculation of the magnetic exchange coupling constants in the austenitic and martensitic phases of Ni-Cu-Mn-Sn. The formation of non-stoichiometric compositions was performed in the coherent potential approximation. Brillouin zone integration was performed using a special point method on a $57 \times 57 \times 57$ k-grid

(4495 k-points) to calculate self-consistent potentials and exchange interaction integrals, respectively. The energy convergence threshold was set to 0.01 mRy.

The temperature dependencies of the magnetization were modeled using the MC method, the Heisenberg Hamiltonian and Metropolis algorithm [31]. The model lattice with periodic boundary conditions consisted of 5488 atoms (for Ni_2MnSn: 2744 Ni atoms, 1372 Mn and Ga atoms) and was obtained by multiplying the 16-atom unit cell $7 \times 7 \times 7$ times. To form $Ni_{50-x}Cu_xMn_{35.94}Sn_{14.06}$ (x = 1.56, 3.12 and 4.69 at.%), the Mn_2 and Cu atoms were randomly distributed at Sn and Ni sites. The number of MC steps per temperature value was 1×10^5. To achieve thermal equilibrium in the system, the first 10^4 steps of the MC were discarded. Magnetization averaging was performed for 225 configurations for every 400 MC steps.

3. Results

3.1. Indirect MCE Estimation

The isothermal change of magnetic entropy ΔS_{iso} was obtained from Maxwell's thermodynamic relations (1) using magnetization curves with increasing field from 0 to 10 T, and is presented in Figure 2b.

$$\Delta S_{iso} = \int_0^H \left(\frac{\partial M(T, H)}{\partial T} \right)_H dH. \quad (1)$$

The maximum value of the magnetic entropy change was ΔS_{iso} = 9.5 J/(kg K) at T = 113 K in $\mu_0 H$ = 10 T (Figure 3a). The ΔS_{iso} values obtained in the vicinity of the hysteresis of the magnetostructural PT are irreversible because reversible MCE values in this region can be obtained by demagnetizing the sample or by turning the magnetic field on again.

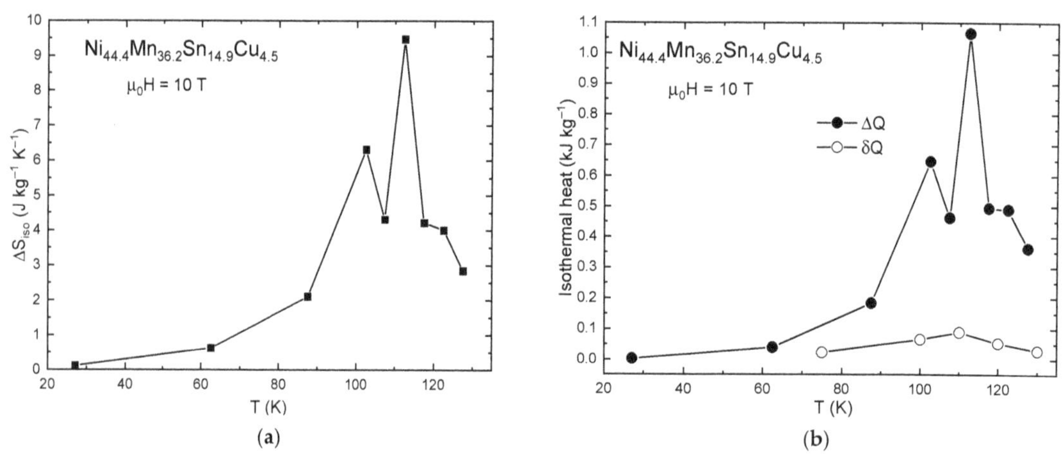

Figure 3. (**a**) Temperature dependence of the isothermal entropy changes ΔS_{iso} in the magnetic field $\mu_0 H$ = 10 T for the Heusler $Ni_{44.4}Mn_{36.2}Sn_{14.9}Cu_{4.5}$ alloy. (**b**) Temperature dependences of the isothermal heat ΔQ and the irreversible magnetization losses δQ in $\mu_0 H$ = 10 T for the Heusler $Ni_{44.4}Mn_{36.2}Sn_{14.9}Cu_{4.5}$ alloy.

The isothermal heat ΔQ is also an important parameter of the working body of a magnetic refrigerator; it allows one to estimate the amount of heat that can be taken from a cooled bath as a result of one ideal cycle of isothermal magnetization/demagnetization processes. Figure 3b shows the ΔQ values calculated from the ΔS_{iso} values using Equation (2):

$$\Delta Q = -T\Delta S_{iso}. \quad (2)$$

The maximum value was ΔQ = 1.06 kJ/kg at T = 113 K in a magnetic field of 10 T, which is an order of magnitude lower than the maximum known value for MnAs at room temperature in the same magnetic field [32].

The presence of field hysteresis (Figure 2a) in the vicinity of the magnetostructural PT in the sample's magnetization/demagnetization cycle leads to irreversible heat release δQ, which was calculated using Equation (3):

$$\partial Q = \mu_0 \oint HdM. \quad (3)$$

Figure 3b shows the δQ values calculated for different temperatures with the maximal value δQ = 0.09 kJ/kg at T = 110 K. This value is only 8.5% of the maximal ΔQ value, which makes the Heusler $Ni_{44.4}Mn_{36.2}Sn_{14.9}Cu_{4.5}$ alloy interesting for use as a magnetocaloric working body in the temperature range of natural gas liquefaction [2], despite the presence of field hysteresis in the magnetostructural PT.

3.2. Direct MCE Measurements

The extraction method in a Bitter magnet field of up to 10 T was used for direct measurement of the MCE in the Heusler $Ni_{44.4}Mn_{36.2}Sn_{14.9}Cu_{4.5}$ alloy under adiabatic conditions ΔT_{ad}. This method was described in detail in [33]. The Bitter magnet produced a steady magnetic field. The temperature change of the sample due to MCE was measured using a differential chromel–gold thermocouple with an accuracy of 0.05 K. The temperature was controlled using a Lake Shore thermo-controller. The sample was placed within a thermal screen to minimize heat exchange with the environment. Movement of the sample to and from the region of the maximum magnetic field was forced with the help of a special actuator. The sample was moved for a duration of 1 s from the center of the Bitter magnet to the outside (or vice versa) at a distance of 35 cm. It should be noted that the magnetic field values before magnetization and after demagnetization were less than 2% of the maximum field in the center of the Bitter coil.

The studies were carried out in two different regimes: (1) sequential heating of the sample, (2) thermal cycling of the sample. In the first case, the sample temperature was successively raised and ΔT_{ad} was measured at selected temperatures. In the second case, the sample was pre-cooled to 4.2 K, and then its temperature was raised to the required temperature for measurement. Figure 4 shows the results of measurements of the ΔT_{ad} value in magnetic fields of 1.8 T and 10 T: closed circles indicate sequential heating and open circles indicate thermal cycling. The inverse MCE was observed in the low-temperature region in a field of 1.8 T in both measurement regimes (Figure 4). The maximum value of the inverse MCE in a field of 1.8 T was observed in the vicinity of the 1st order PT, with a value of ΔT_{ad} = –0.5 K at an initial temperature T_0 = 117 K (Figure 4). The inverse MCE increased with increasing magnetic field up to 10 T, reaching ΔT_{ad} = –2.7 K at T_0 = 117 K (Figure 4). It is interesting that direct MCE in a magnetic field of 10 T was observed in the FiM martensite phase and reached ΔT_{ad} = 0.7 K (Figure 4) in the sequential heating regime in the temperature range of 20–60 K.

3.3. Computational Results

We will now discuss the results of the geometric optimization of the crystal structures of the austenitic and martensitic phases as well as the calculation of the magnetic properties of Heusler $Ni_{50-x}Cu_xMn_{35.94}Sn_{14.06}$ alloys.

Figure 5 shows the energy landscape $E(c/a)$ of $Ni_{50-x}Cu_xMn_{35.94}Sn_{14.06}$, including the response to volume-conserving elongations and compressions of the cubic $L2_1$ structure along the c axis. The results are shown for both FM and FiM solutions (Figure 5a). It can be seen that the FM solution exhibits only one global cubic minimum at c/a = 1, which is 11 meV/atom higher in energy than observed for FiM, indicating an instability of the FM tetragonal phase. For all compounds, the FiM ordering was energetically preferable to the FM ordering for both cubic austenitic and tetragonal martensitic phases. The global minimum for the FiM tetragonal phase took place at around c/a = 1.25.

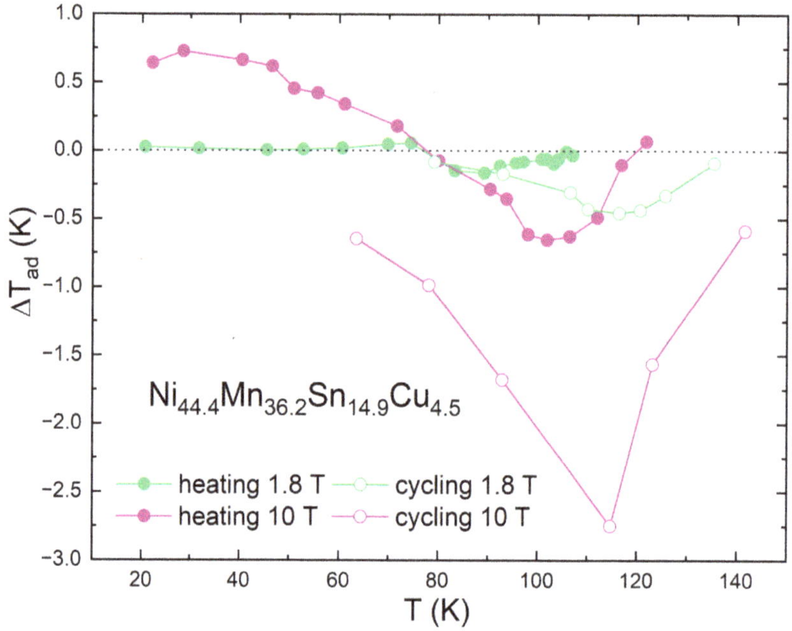

Figure 4. Temperature dependences of the adiabatic temperature change ΔT_{ad} for the $Ni_{44.4}Mn_{36.2}Sn_{14.9}Cu_{4.5}$ Heusler alloy, measured in magnetic fields of 1.8 T (green curves) and 10 T (violet curves) in two different regimes: solid circles—during sequential heating of the sample; open circles—after pre-cooling down to 4.2 K and subsequent heating to the measurement temperature.

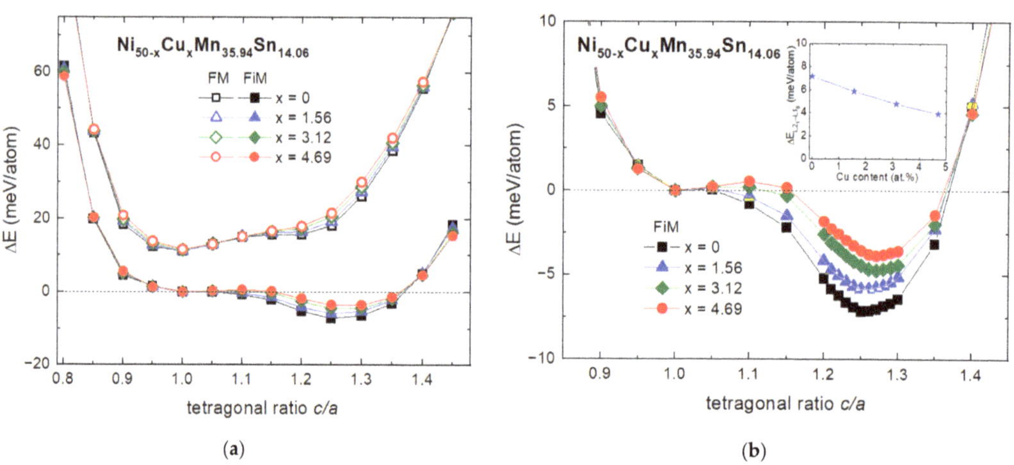

Figure 5. (**a**) Energy difference as a function of the tetragonal c/a ratio for $Ni_{50-x}Cu_xMn_{35.94}Sn_{14.06}$ (x = 0, 1.56, 3.12, and 4.69). For all compounds, the zero energy corresponds to the FiM cubic structure. Filled and open symbols denote the FiM and FM solutions, respectively. (**b**) Enlarged scale for the FiM solution. The inset shows the energy difference between FiM cubic and tetragonal phases as a function of Cu content.

A close look at the FiM $E(c/a)$ curves shown in Figure 5b reveals that the c/a ratio of the tetragonal phase varied slightly from 1.25 to 1.27 with increasing Cu content due to a reduction (increase) in the tetragonal lattice constant a_t (c), respectively (see Table 1). In addition, the optimized cubic lattice constant increased slightly with the addition of Cu. This is explained by the slightly larger atomic radius of Cu (r = 1.28 Å) compared to that of Ni (r = 1.24 Å). It should be noted that partial substitution of Ni with Cu revealed an almost linear decrease in the energy difference ($\Delta E_{L2_1-L1_0}$) between the FIM cubic and tetragonal structure, as is evident from Figure 5b. For the parent compound with x = 0, $\Delta E_{L2_1-L1_0}$ was calculated to be 7.145 meV/atom, whereas in the case of x = 4.69 at.%, it reduced to 3.89 meV/atom. The decrease in the energy barrier with the Cu content indirectly indicates the reduction in the martensitic transition temperature T_m. The T_m temperature can be estimated from a rough approximation: $\Delta E_{L2_1-L1_0} \approx k_B T_m$, where k_B is the Boltzmann constant. According to this expression, $T_m \approx$ 83 K for $Ni_{50}Mn_{35.94}Sn_{14.06}$ and 45 K for $Ni_{45.31}Cu_{4.69}Mn_{35.94}Sn_{14.06}$.

Table 1. Optimized lattice parameters and Debye temperatures of cubic and tetragonal structures of $Ni_{50-x}Cu_xMn_{35.94}Sn_{14.06}$. The Debye temperatures were calculated from elastic moduli.

Phase	Parameters	x = 0	x = 1.56	x = 3.12	x = 4.69
FiM cubic (c/a = 1)	a_0 (Å)	5.953	5.955	5.959	5.963
	Θ_D (K)	316	308	293	282
FiM tetragonal (c/a = 1.27)	a_t (Å)	5.526	5.513	5.502	5.500
	c (Å)	6.908	6.947	6.988	6.994
	c/a	1.25	1.26	1.27	1.27
	Θ_D (K)	333	325	308	295

A more correct way to estimate the martensitic transition temperature is to calculate the free energies of the austenitic and martensitic phases. For simplicity of calculation, we only considered the lattice contribution to the free energy, which at low temperatures plays a predominant role compared to the electronic and magnetic contributions:

$$F(T,V) = E(V) + F_{lat}(T,V), \quad (4)$$

where $E(V)$ is the ground-state energy calculated at T = 0 K, and $F_{lat}(T,V)$ is the lattice contribution calculated within the Debye model [34].

In Figure 6, we illustrate the free energies of $L2_1$ cubic and $L1_0$ tetragonal phases as well as the free-energy difference $\Delta F = F_{L2_1} - F_{L1_0}$ for the $Ni_{45.31}Cu_{4.69}Mn_{35.94}Sn_{14.06}$ alloy as an example. For $\Delta F > 0$, the $L1_0$ phase (martensite) is preferable and, vice versa, the $L2_1$ phase (austenite) is stable for $\Delta F < 0$. As is evident from the figure, both free-energy curves reveal a non-linear behavior with temperature and intersect with each other at low temperatures. The martensitic transition temperature extracted from $\Delta F = 0$ is about 88 K, which agrees well with the experimental value (85.75 K) for $Ni_{44.4}Mn_{36.2}Sn_{14.9}Cu_{4.5}$, where T_m was computed via $T_m = (M_s + M_f + A_f + A_s)/4$. The low T_m temperature suggests that the zero-point vibrational energy ($\frac{9}{8}k_B\Theta_D$) plays a predominant role as compared to a vibrational entropy in the F_{lat} term. Thus, T_m is mainly contributed by $\Delta E_{L2_1-L1_0}$ at T = 0 K and $\Delta\Theta_D^{L2_1-L1_0}$ at T > 0 K. For $Ni_{45.31}Cu_{4.69}Mn_{35.94}Sn_{14.06}$, $\Delta\Theta_D^{L2_1-L1_0}$ is 13 K in absolute value. We would like to note that T_m is sensitive to $\Delta\Theta_D^{L2_1-L1_0}$, and an increase in $\Delta\Theta_D^{L2_1-L1_0}$ ($\Theta_D^{L2_1} < \Theta_D^{L1_0}$) leads to a reduction in T_m; the austenitic phase becomes stable in the whole temperature range at a $\Delta\Theta_D^{L2_1-L1_0}$ of about 30 K in absolute value. On the other hand, an increase in $\Delta\Theta_D^{L2_1-L1_0}$ ($\Theta_D^{L2_1} > \Theta_D^{L1_0}$) shifts T_m to higher temperatures.

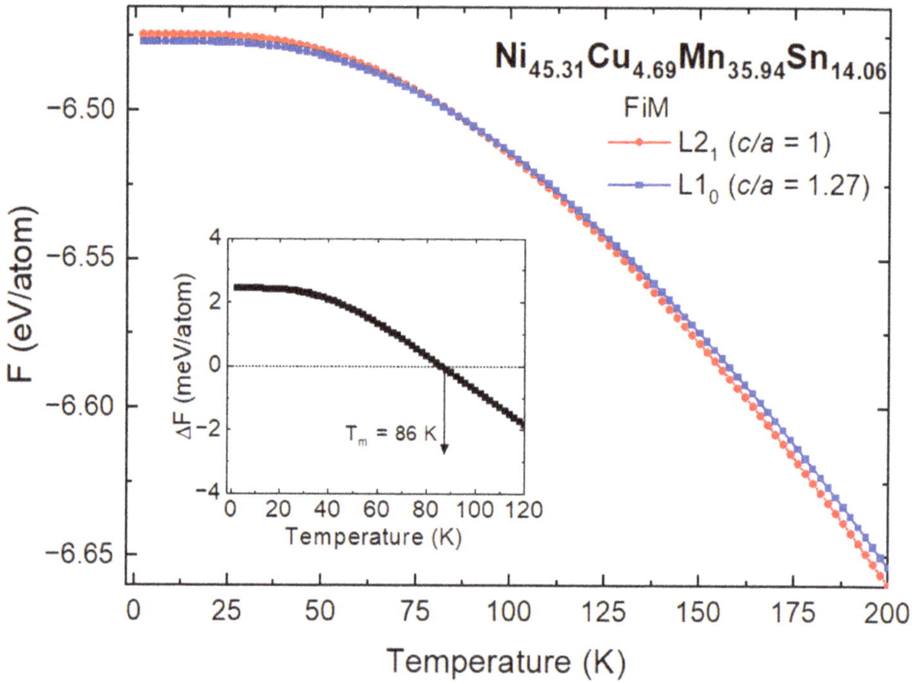

Figure 6. Temperature dependencies of the free energy of cubic and tetragonal phases of $Ni_{45.31}Cu_{4.69}Mn_{35.94}Sn_{14.06}$. The inset shows the free-energy difference between $L2_1$ and $L1_0$ structures.

Figure 7 shows the pairwise exchange coupling constants J_{ij} as function of the distance between *i* and *j* atoms for FiM-$L2_1$ and FiM-$L1_0$ $Ni_{45.31}Cu_{4.69}Mn_{35.94}Sn_{14.06}$ alloy as an example. The magnetic interactions reveal a similar behavior for all compounds under study because Cu atoms are nonmagnetic. For both phases, intra-sublattice J_{ij} constants between Mn atoms (Mn_1-Mn_1 and Mn_2-Mn_2) show similar damped oscillatory behavior as a function of the distance between atoms up to $d/a \approx 2$, with the exception of the interactions for the three coordination shells of cubic structure. In this case, J_{ij} (Mn_1-Mn_1) demonstrates a reducing FM character, whereas J_{ij} (Mn_2-Mn_2) within the first and second coordination shells exhibits a reducing AFM character.

The cubic inter-sublattice interactions between the nearest Mn_1-Mn_2 atoms located at a smaller distance ($d/a = 0.5$) compared to the nearest pairs $Mn_{1(2)}$-$Mn_{1(2)}$ ($d/a = \sqrt{2}/2$) are characterized by a strong AFM interaction, which gradually decreases with increasing d/a. In contrast to the cubic structure, the strongest AFM exchange ($\approx$–20 meV) between the four nearest atoms Mn_1-Mn_2 located in the (110) plane is found for the tetragonal structure due to the smallest distance ($d/a = 0.5$), whereas the two next nearest atoms Mn_1-Mn_2 ($d/a \approx 0.627$) exhibit a strong ferromagnetic interaction with $J_{ij} \approx 11.5$ meV. The $Mn_{1(2)}$-Ni interactions show a similar behavior for both phases, demonstrating the FM exchange only between the nearest neighbors. Generally, the behavior of J_{ij} coupling constants is similar to those that were previously reported for Ni-Mn-Sn alloys [35].

Let us make a general remark concerning the behavior of the exchange interaction parameters for the remaining compounds in the cubic and tetragonal phases. Since the substitution of Ni with Cu influences a small change in the crystal structure parameters for both phases, it will also weakly affect the exchange constants between the nearest pairs of Mn_1–Ni and Mn_2–Ni (see Table 2). For the cubic structure, a slight enhancement of the FM interactions $J_{Mn1(2)\text{-}Mn1(2)}$ and $J_{Mn1(2)\text{-}Ni}$ and a weakening of the AFM interaction $J_{Mn1\text{-}Mn2}$ with increasing Cu content is observed, which also affects the Curie temperature of austenite ($T_C{}^A$). In the case of the tetragonal structure, a weakening of FM and AFM interactions between Ni, Mn_1 and Mn_2 atoms is observed. However, the Curie temperature of martensite, $T_C{}^M$, increases similarly to $T_C{}^A$. The increase in $T_C{}^M$ is caused by the weakening of the strong AFM interaction between the nearest Mn_1-Mn_2, the change of which has a larger contribution to the magnetic energy compared to the change of $J_{Mn1(2)\text{-}Mn1(2)}$ and $J_{Mn1(2)\text{-}Ni}$ interactions as a function of Cu content.

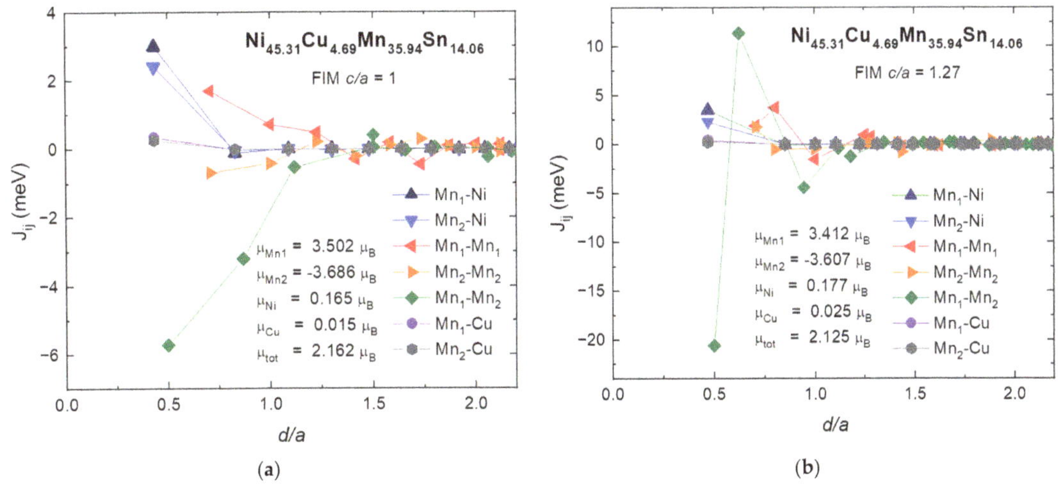

Figure 7. Calculated magnetic exchange parameters of (**a**) FiM $L2_1$- and (**b**) $L1_0$-$Ni_{45.31}Cu_{4.69}Mn_{35.94}Sn_{14.06}$ as functions of the distance between the atoms in units of the lattice constant *a*. Positive J_{ij} values denote FM interactions, and negative ones denote AFM interactions.

Table 2. The nearest exchange coupling constants and Curie temperatures calculated within the MC method and MFA for cubic and tetragonal structures of Heusler $Ni_{50-x}Cu_xMn_{35.94}Sn_{14.06}$ alloys.

x	Mn_1-Ni	Mn_2-Ni	Mn_1-Mn_1	Mn_2-Mn_2	Mn_1-Mn_2	T_C (MC)	T_C (MFA)
			FiM cubic structure				
0	2.961	2.341	1.082	−0.936	−6.065	425	423.4
1.56	2.977	2.364	1.201	−0.837	−5.925	430	433.5
3.12	2.988	2.384	1.492	−0.772	−5.859	452	444.8
4.69	2.999	2.402	1.702	−0.678	−5.722	464	453.1
			FiM tetragonal structure				
0	3.514	2.308	2.035	1.859	−21.902	484	544.9
1.56	3.502	2.299	1.985	1.826	−21.513	506	547.5
3.12	3.495	2.294	1.942	1.798	−21.071	514	549.6
4.69	3.489	2.287	1.906	1.779	−20.598	523	551.4

Figure 8a shows the T–x phase diagram for $Ni_{50-x}Cu_xMn_{35.94}Sn_{14.06}$, including the predicted T_m and T_C temperatures and the experimentally measured values of T_m and $T_C{}^A$. With increasing Cu doping level, T_m reduces nonlinearly from 133 K (x = 0 at.%) to 88 K (x = 4.69 at.%). This is mainly due to a decrease in the energy barrier $\Delta E_{L2_1-L1_0}$ and $\Delta\Theta_D^{L2_1-L1_0}$, both of which have maxima at x = 0 (see Figure 5 and Table 1). The predicted value of T_m for $Ni_{45.31}Cu_{4.69}Mn_{35.94}Sn_{14.06}$ is close to the experimental one for $Ni_{44.4}Cu_{4.5}Mn_{36.2}Sn_{14.9}$. As for the Curie temperature, MC simulations and mean-field approximation (MFA) show a similar $T_C{}^A$ trend with the increase in Cu content. Nevertheless, the experimental value of $T_C{}^A$ for $Ni_{44.4}Cu_{4.5}Mn_{36.2}Sn_{14.9}$ is less than the theoretical one by about 100 K.

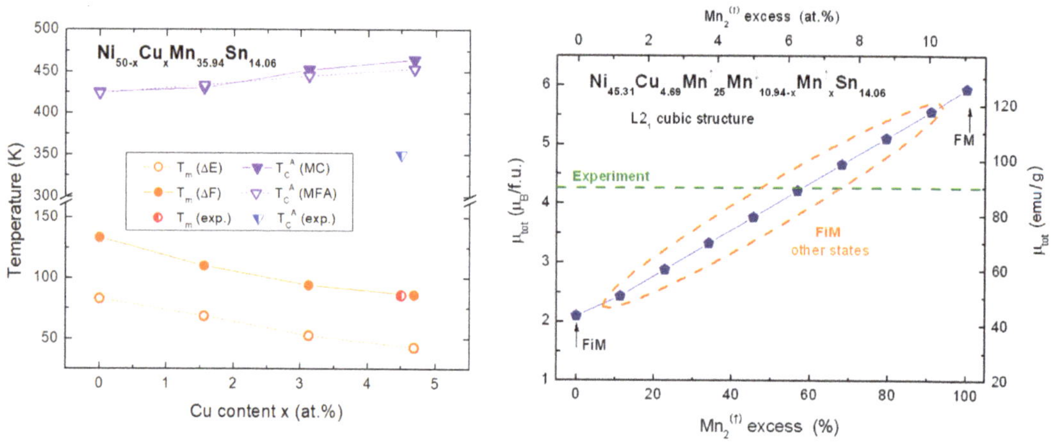

(**a**) (**b**)

Figure 8. (**a**) Calculated phase diagram for $Ni_{50-x}Cu_xMn_{35.94}Sn_{14.06}$ (x = 0, 1.56, 3.12 and 4.69). Here, T_m (ΔE) and T_m (ΔF) are the martensitic transformation temperatures calculated from the energy difference at T = 0 K, which has been converted to a temperature scale, and from the free-energy difference considering only lattice contribution, respectively; $T_C{}^A$ is the Curie temperatures of austenite calculated by MC and MFA methods. The experimental values of T_m and $T_C{}^A$ refer to $Ni_{44.4}Mn_{36.2}Sn_{14.9}Cu_{4.5}$, and T_m is computed via $T_m = (M_s + M_f + A_f + A_s)/4$. (**b**) The calculated total magnetic moment (filled symbols) of $Ni_{45.31}Cu_{4.69}Mn_{35.94}Sn_{14.06}$ with the cubic structure as a function of Mn_2 atoms at Sn sites with a parallel magnetic moment to that of Mn_1 atoms. The experimental magnetization for the austenitic phase of $Ni_{44.4}Mn_{36.2}Sn_{14.9}Cu_{4.5}$ is highlighted with the green dashed line.

4. Discussion

Let us next discuss the calculated total magnetic moment μ_{tot} for $Ni_{45.31}Cu_{4.69}Mn_{35.94}Sn_{14.06}$ and its relation to the experimentally obtained moment for $Ni_{44.4}Cu_{4.5}Mn_{36.2}Sn_{14.9}$. It follows from Figure 6 that the magnetic reference state for both austenite and martensite is an FiM one, where all magnetic moments of $Mn_1^{\uparrow}$ and $Mn_2^{\downarrow}$ are oppositely aligned. As a consequence, the μ_{tot} is calculated to be almost equal 2.16 and 2.12 μ_B or 45.6 and 44.8 emu/g, respectively. The latter value is close to the experimental magnetization of martensite (≈53 emu/g). However, there is some discrepancy between the calculated and measured values of μ_{tot} for austenite. The experiment yielded a μ_{tot} value that was double the calculated value (90 emu/g—dashed line in Figure 8b). On the other hand, the calculated value of μ_{tot} for FM austenite, which is energetically unfavorable, is about 5.94μ_B or 125.6 emu/g larger than the experimental one. The dashed line in Figure 2a is the calculated curve (approximation) of the temperature dependence of the magnetization in $\mu_0 H$ = 3 T for the sample in the austenite phase without a magnetostructural PT. This curve was calculated using the approach proposed in [36]. The magnetization value obtained using this approach is 110 emu/g (Figure 2a). This is the estimation of the maximal magnetization of the austenitic phase "without martensitic transition", and it is very close to the calculated value.

To explain the observed discrepancy, we suggest that not all Mn_2 atoms located at the Sn sublattice have an antiparallel magnetic moment with respect to Mn_1 atoms located at the regular Mn sublattice. To prove this, we additionally performed CPA calculations of μ_{tot} for $L2_1$-$Ni_{45.31}Cu_{4.69}Mn_{35.94}Sn_{14.06}$ (or $Ni_{45.31}Cu_{4.69}Mn_{25}^{\uparrow}Mn_{10.94-x}^{\downarrow}Mn_x^{\uparrow}Sn_{14.06}$) using SPR-KKR code as a function of Mn_2 atoms at Sn sites with a parallel magnetic moment to that of Mn_1 atoms, as shown in Figure 8b. As can be seen from the figure, μ_{tot} increases linearly with the fraction of $Mn_2^{\uparrow}$ atoms and reaches 5.94 μ_B or 125.6 emu/g for 100% of the

$Mn_2^{\uparrow}$ spins aligned parallel to the $Mn_1^{\uparrow}$ ones (FM order). We found that the experimental value of μ_{tot} can be reached theoretically if only $\approx$40% of the Mn_2 atoms interact antiferromagnetically with the Mn_1 atoms. This finding allows us to conclude that the experimental sample had some kind of FiM order, with a predominance of FM-ordered Mn atoms.

The inverse MCE at low temperatures in the Heusler $Ni_{33.7}Co_{14.8}Mn_{35.4}Ti_{16.1}$ alloy showed similar values at low temperatures [23]: maximal ΔT_{ad} = −2.7 K at T_0 = 90 K in a magnetic field of 20 T. This may be a physical limit on the value of the inverse MCE for Heusler alloys with PTs at low temperatures, which is significantly lower than in known Heusler alloys with the inverse MCE at room temperature, for example [37]. The further study of medium-entropy [38–40] and high-entropy [41–43] alloys that do not contain rare-earth elements will reveal compositions with high inverse MCE for application in low-temperature SMC.

5. Conclusions

The inverse MCE in a Heusler $Ni_{44.4}Mn_{36.2}Sn_{14.9}Cu_{4.5}$ alloy in the low-temperature region was observed in a magnetic field of 1.8 T in both measurement regimes: sequential heating and thermal cycling. The maximum value of the inverse MCE in the vicinity of the first-order PT in a magnetic field of 1.8 T was observed to be ΔT_{ad} = −0.5 K at an initial temperature T_0 = 117 K. The inverse MCE reached up to ΔT_{ad} = −2.7 K at T_0 = 117 K with the increase in the magnetic field up to 10 T. The direct MCE in the FiM martensite phase was observed in a magnetic field of 10 T and reached the maximal value of ΔT_{ad} = 0.7 K in the sequential heating regime in the temperature range of 20–60 K.

First-principles and MC approaches were applied to determine the phase diagram for FM-ordered $Ni_{50-x}Cu_xMn_{35.94}Sn_{14.06}$ (x = 0, 1.56, 3.12 and 4.69) compounds, which were similar in composition to the experimental sample. Cu doping led to a decrease in the T_m temperature and an increase in the $T_C{}^A$ temperature. This was mainly due to the slight change in the optimized lattice constants, magnetic exchange interactions and Debye temperatures for austenite and martensite. From the analysis of our computational results, we suggest that the experimental sample exhibited some type of FiM order in the austenite, with a predominance of ferromagnetically ordered Mn atoms. This theory was satisfactorily confirmed by our experiment, which will make it possible to predict the transition temperatures, magnetizations and MCE properties of Heusler Ni-Mn-Sn family alloys when a portion of the Sn atoms are replaced with Cu.

Author Contributions: Conceptualization, A.P.K. and V.G.S.; Methodology, Y.S.K. and V.D.B.; Software, V.V.S.; Formal analysis, A.P.K., Y.S.K. and V.V.S.; Investigation, Y.S.K., A.V.K., R.Y.G., I.I.M. and V.V.S.; Resources, A.V.M. and J.Ć.; Writing—original draft, A.P.K., Y.S.K. and V.V.S.; Supervision, V.D.B. and V.G.S.; Project administration, V.G.S. All authors of the manuscript contributed equally to data acquisition and analysis, writing, editing and formatting. All authors have read and agreed to the published version of the manuscript.

Funding: The magnetization and magnetocaloric measurements were carried out within the framework of Russian Science Foundation project № 20-19-00745, https://rscf.ru/project/20-19-00745/ (accessed on 1 December 2023). The alloy preparation and structure analysis were carried out by R.Y.G and I.I.M. within the framework of the state task of IMSP RAS. V.V.S. the financial support from the Priority-2030 Program of NUST "MISiS" (grant No. K2-2022-022) (the total energy calculations). V.D.B. acknowledges the Ministry of Science and Higher Education of the Russian Federation within the Russian State Assignment, under Contract 075-01493-23-00 (the phonon calculations).

Data Availability Statement: The data presented in this study are openly available in Preprints.org at https://doi.org/10.20944/preprints202310.0715.v1.

Conflicts of Interest: The authors declare no conflict of interest.

References

1. Giauque, W.F.; MacDougall, D.P. Attainment of Temperatures Below 1 Absolute by Demagnetization of $Gd_2(SO_4)_3 \cdot 8H_2O$. *Phys. Rev.* **1933**, *43*, 768. [CrossRef]
2. Koshkid'ko, Y.S.; Dilmieva, E.T.; Kamantsev, A.P.; Mashirov, A.V.; Cwik, J.; Kol'chugina, N.B.; Koledov, V.V.; Shavrov, V.G. Magnetocaloric Materials for Low-Temperature Magnetic Cooling. *J. Commun. Technol. Electron.* **2023**, *68*, 379–388. [CrossRef]
3. Tishin, A.M.; Spichkin, Y.I. *The Magnetocaloric Effect and Its Applications*; IOP Publisher: Bristol, UK, 2003; p. 476.
4. Belov, K.P. *The Magnetothermal Phenomena in Rare-Earth Magnetics*; Nauka: Moscow, Russia, 1990.
5. Khovaylo, V.V.; Taskaev, S.V. Magnetic Refrigeration: From Theory to Applications. *Encycl. Smart Mater.* **2022**, *5*, 407–417.
6. Tang, X.; Sepehri-Amin, H.; Terada, N.; Martin-Cid, A.; Kurniawan, I.; Kobayashi, S.; Kotani, Y.; Takeya, H.; Lai, J.; Matsushita, Y.; et al. Magnetic refrigeration material operating at a full temperature range required for hydrogen liquefaction. *Nat. Commun.* **2022**, *13*, 1817. [CrossRef] [PubMed]
7. Aziz, M. Liquid Hydrogen: A Review on Liquefaction, Storage, Transportation, and Safety. *Energies* **2021**, *14*, 5917. [CrossRef]
8. Zhang, H.; Sun, Y.J.; Niu, E.; Yang, L.H.; Shen, J.; Hu, F.X.; Sun, J.R.; Shen, B.G. Large magnetocaloric effects of RFeSi (R = Tb and Dy) compounds for magnetic refrigeration in nitrogen and natural gas liquefaction. *Appl. Phys. Lett.* **2013**, *103*, 202412. [CrossRef]
9. Timmerhaus, K.D.; Reed, R.P. *Cryogenic Engineering: Fifty Years of Progress*; Springer Science & Business Media: New York, NY, USA, 2007.
10. Andreenko, A.S.; Belov, K.P.; Nikitin, S.A.; Tishin, A.M. Magnetocaloric effects in rare-earth magnetic materials. *Sov. Phys. Usp.* **1989**, *32*, 649–664. [CrossRef]
11. Zheng, X.Q.; Xu, Z.Y.; Zhang, B.; Hu, F.X.; Shen, B.G. The normal and inverse magnetocaloric effect in RCu_2 (RB = Tb, Dy, Ho, Er) compounds. *J. Magn. Magn. Mater.* **2017**, *421*, 448–452. [CrossRef]
12. Kuchin, A.G.; Iwasieczko, W.; Platonov, S.P. The magnetocaloric effect in R_2Fe_{17} intermetallics with different types of magnetic phase transition. *Low Temp. Phys.* **2015**, *41*, 985–989. [CrossRef]
13. Kamantsev, A.P.; Koshkidko, Y.S.; Taskaev, S.V.; Khovaylo, V.V.; Koshelev, A.V.; Cwik, J.; Shavrov, V.G. Inverse Magnetocaloric Effect and Kinetic Arrest Behavior in As-Cast Gd_2In at Cryogenic Temperatures. *J. Supercond. Nov. Magn.* **2022**, *35*, 2181–2186. [CrossRef]
14. Kuchin, A.G.; Platonov, S.P.; Mukhachev, R.D.; Lukoyanov, A.V.; Volegov, A.S.; Gaviko, V.S.; Yakovleva, M.Y. Large Magnetic Entropy Change in GdRuSi Optimal for Magnetocaloric Liquefaction of Nitrogen. *Metals* **2023**, *13*, 290. [CrossRef]
15. Nikitin, S.A.; Skokov, K.P.; Koshkid'ko, Y.S.; Pastushenkov, Y.G.; Ivanova, T.I. Giant Rotating Magnetocaloric Effect in the Region of Spin-Reorientation Transition in the $NdCo_5$ Single Crystal. *Phys. Rev. Lett.* **2010**, *105*, 137205. [CrossRef] [PubMed]
16. Moon, J.Y.; Kim, M.K.; Oh, D.G.; Kim, J.H.; Shin, H.J.; Choi, Y.J.; Lee, N. Anisotropic magnetic properties and giant rotating magnetocaloric effect in double-perovskite Tb_2CoMnO_6. *Phys. Rev. B* **2018**, *98*, 174424. [CrossRef]
17. Koshkid'ko, Y.; Cwik, J.; Dilmieva, E.; Pandey, S.; Quetz, A.; Aryal, A.; Dubenko, I.; Ali, N.; Granovsky, A.; Lähderanta, E.; et al. Inverse magnetocaloric effects in metamagnetic Ni-Mn-In-based alloys in high magnetic fields. *J. Alloys Compd.* **2017**, *695*, 3348–3352. [CrossRef]
18. Konoplyuk, S.M.; Kokorin, V.V.; Mashirov, A.V.; Kamantsev, A.P.; Koledov, V.V.; Shavrov, V.G.; Koshelev, A.V. Direct measurements of adiabatic temperature change in $Ni_{49.9}Mn_{37.03}Sb_{12.3}Fe_{0.77}$ alloy due to magnetocaloric effect in the temperature range of martensitic transformation. *IEEE Trans. Magn.* **2018**, *54*, 2500204. [CrossRef]
19. Sokolovskiy, V.V.; Nachinova, D.V.; Buchel'nikov, V.D.; Dilmieva, E.T.; Koshkidko, Y.S.; Emelyanova, S.M.; Marchenkova, E.B.; Marchenkov, V.V. Magnetic and magnetocaloric properies of Heusler alloys Ni-Mn-Sn with an excess of Mn within the theoretica-land experimental approaches. *Chel. Phys.-Math. J.* **2020**, *5*, 493–503. [CrossRef]
20. Lekkla, P.; Jantaratana, P.; Chotibhawaris, T. Magnetocaloric Effect in CoFe-Electroplated $Ni_{50}Mn_{33}In_{16}Cr_1$ Alloy. *Metals* **2022**, *12*, 2137. [CrossRef]
21. Dadda, K.; Alleg, S.; Souilah, S.; Daza, J.; Saurina, J.; Suñol, J.-J.; Bessais, L.; Hlil, E.-K. Microstructure, Critical Behavior and Magnetocaloric Properties of Melt-Spun $Ni_{51.82}Mn_{32.37}In_{15.81}$. *Magnetochemistry* **2022**, *8*, 179. [CrossRef]
22. Marchenkov, V.V.; Emelyanova, S.M.; Marchenkova, E.B. Martensitic Transformation Temperatures and Hall Effect in $Ni_{47-x}Mn_{41+x}In_{12}$ (x = 0, 1, 2) Alloys. *Materials* **2023**, *16*, 672. [CrossRef]
23. Beckmann, B.; Koch, D.; Pfeuffer, L.; Gottschall, T.; Taubel, A.; Adabifiroozjaei, E.; Miroshkina, O.N.; Riegg, S.; Niehoff, T.; Kani, N.A.; et al. Dissipation losses limiting first-order phase transition materials in cryogenic caloric cooling: A case study on all-d-metal Ni (-Co)-Mn-Ti Heusler alloys. *Acta Mater.* **2023**, *246*, 118695. [CrossRef]
24. Dincer, I.; Yüzüak, E.; Elerman, Y. Influence of irreversibility on inverse magnetocaloric and magnetoresistance properties of the $(Ni, Cu)_{50}Mn_{36}Sn_{14}$ alloys. *J. Alloys Compd.* **2010**, *506*, 508–512. [CrossRef]
25. Gaifullin, R.Y.; Kirilyuk, K.K.; Safarov, I.M.; Musabirov, I.I. Structure of $Ni_{44.4}Mn_{36.2}Sn_{14.9}Cu_{4.5}$ alloy applicable for thermomechanical treatment. *Lett. Mater.* **2023**, *13*, 164–170. [CrossRef]
26. Nizhankovskii, V.I.; Lugansky, L.B. Vibrating sample magnetometer with a step motor. *Meas. Sci. Technol.* **2007**, *18*, 1533. [CrossRef]
27. Kresse, G.; Furthmüller, J. Efficient iterative schemes for ab initio total-energy calculations using a plane- wave basis set. *Phys. Rev. B* **1996**, *54*, 11169–11186. [CrossRef]
28. Kresse, G.; Joubert, D. From ultrasoft pseudopo-tentials to the projector augmented-wave method. *Phys. Rev. B* **1999**, *59*, 1758–1775. [CrossRef]

29. Perdew, J.P.; Burke, K.; Ernzerhof, M. Generalized Gradient Approximation Made Simple. *Phys. Rev. Lett.* **1996**, *77*, 3865–3868. [CrossRef]
30. Ebert, H.; Ködderitzsch, D.; Minár, J. Calculating condensed matter properties using the KKR-Green's function method-recent developments and applications. *Rep. Prog. Phys.* **2011**, *74*, 096501. [CrossRef]
31. Landau, D.; Binder, K. *A Guide to Monte Carlo Simulations in Statistical Physics*; Cambridge University Press: Cambridge, UK, 2021; p. 578.
32. Koshkid'ko, Y.S.; Dilmieva, E.T.; Cwik, J.; Rogacki, K.; Kowalska, D.; Kamantsev, A.P.; Koledov, V.V.; Mashirov, A.V.; Shavrov, V.G.; Valkov, V.I.; et al. Giant reversible adiabatic temperature change and isothermal heat transfer of MnAs single crystals studied by direct method in high magnetic fields. *J. Alloys Comp.* **2019**, *798*, 810–819. [CrossRef]
33. Koshkid'ko, Y.S.; Ćwik, J.; Ivanova, T.I.; Nikitin, S.A.; Miller, M.; Rogacki, K. Magnetocaloric properties of Gd in fields up to 14 T. *J. Magn. Magn. Mater.* **2017**, *433*, 234–238. [CrossRef]
34. Moruzzi, V.L.; Janak, J.F.; Schwarz, K. Calculated thermal properties of metals. *Phys. Rev. B* **1988**, *37*, 790–799. [CrossRef]
35. Sokolovskiy, V.V.; Buchelnikov, V.D.; Zagrebin, M.A.; Entel, P.; Sahool, S.; Ogura, M. First-principles investigation of chemical and structural disorder in magnetic $Ni_2Mn_{1+x}Sn_{1-x}$ Heusler alloys. *Phys. Rev. B* **2012**, *86*, 134418. [CrossRef]
36. Kuz'min, M.D. Landau-type parametrization of the equation of state of a ferromagnet. *Phys. Rev. B* **2008**, *77*, 184431. [CrossRef]
37. Gottschall, T.; Skokov, K.P.; Frincu, B.; Gutfleisch, O. Large reversible magnetocaloric effect in Ni-Mn-In-Co. *Appl. Phys. Lett.* **2015**, *106*, 021901. [CrossRef]
38. Timofeeva, E.; Panchenko, E.; Zherdeva, M.; Tokhmetova, A.; Chumlyakov, Y.I. The Cyclic Stability of the Superelasticity in Quenched and Aged $Ni_{44}Fe_{19}Ga_{27}Co_{10}$ Single Crystals. *Metals* **2023**, *13*, 1538. [CrossRef]
39. Bhale, P.; Ari-Gur, P.; Noebe, R.D.; Ren, Y.; Madiligama, A.; Devaraj, R.; Cook, M.S. Effect of Melt-Spinning Parameters on the Structure and Properties of $Ni_{55.5}Mn_{18.8}Ga_{24}Si_{1.7}$ Heusler Alloy Ribbons. *Materials* **2023**, *16*, 6590. [CrossRef] [PubMed]
40. Bachagha, T.; Chakaravarthy, R.; Ren, W.; Saurina, J.; Suñol, J.-J. Structural, Magnetocaloric, and Magnetic Properties in Heusler $Ni_{50}Mn_{35}In_{10}X_5$ (X = Ga, Fe and Al) Alloys. *Metals* **2023**, *13*, 1913. [CrossRef]
41. Zhang, Y.; Hao, W.; Hu, C.; Wang, X.; Zhang, X.; Li, L. Rare-Earth-Free $Mn_{30}Fe_{20-x}Cu_xAl_{50}$ Magnetocaloric Materials with Stable Cubic CsCl-Type Structure for Room-Temperature Refrigeration. *Adv. Funct. Mater.* **2023**, 2310047. [CrossRef]
42. Zhang, Y.; Xu, P.; Zhu, J.; Yan, S.; Zhang, J.; Li, L. The emergence of considerable room temperature magnetocaloric performances in the transition metal high-entropy alloys. *Mater. Today Phys.* **2023**, *32*, 101031. [CrossRef]
43. Nambiar, S.S.; Murthy, B.R.N.; Karthik, B.M.; Sharma, S.; Prasanna, A.A. Investigation on Magnetization, Magnetocalory, Magnetoresistance, and Electric Properties of Ni-Mn Based Heusler Alloy. *J. Compos. Sci.* **2023**, *7*, 5. [CrossRef]

metals

MDPI

Article

The Cyclic Stability of the Superelasticity in Quenched and Aged $Ni_{44}Fe_{19}Ga_{27}Co_{10}$ Single Crystals

Ekaterina Timofeeva *, Elena Panchenko, Maria Zherdeva, Aida Tokhmetova and Yuriy I. Chumlyakov *

Laboratory for Physics of High-Strength Crystals, Siberian Physical-Technical Institute, National Research Tomsk State University, Lenina Str. 36, 634050 Tomsk, Russia; panchenko@mail.tsu.ru (E.P.); zhmv@mail.tsu.ru (M.Z.); aida.tokhmetova@mail.tsu.ru (A.T.)

* Correspondence: katie@mail.tsu.ru (E.T.); chum@phys.tsu.ru (Y.I.C.)

Abstract: The effect of aging (at 773 K for 1 h) on the cyclic stability of superelasticity was studied in preliminarily quenched (after annealing at 1448 K for 1 h) $Ni_{44}Fe_{19}Ga_{27}Co_{10}$ [001]-oriented single crystals. It was shown that NiFeGaCo single crystals (both quenched and aged) exhibited highly stable superelasticity during more than 10^5 loading/unloading cycles in compression without destruction. The degradation of the superelasticity curves was divided into stages, and each stage of degradation was carefully considered. The precipitation of ω-phase particles led to a change in the degradation mechanism of superelasticity in the aged crystals. The main degradation mechanism for quenched crystals is a formation of uniform distribution of dislocations; for aged crystals, the microstructural degradation mechanisms consist of a non-uniform distribution of dislocations around the particles and the appearance of residual martensite.

Keywords: cyclic stability; superelasticity; single crystals; martensitic transformations; critical stress; stress hysteresis; aging

1. Introduction

The Heusler NiFeGaCo alloy is one of the promising shape memory alloys. These alloys demonstrate multistage thermoelastic martensitic transformations (MT) from $B2(L2_1)$-austenite to $L1_0$-martensite through modulated 10M(14M)-phases [1]. Such effects, such as superelasticity (SE) and shape memory effect (SME), are observed on the basis of MT. SE is observed during loading/unloading cycles and is characterized by significant inelastic strain. The given inelastic strain is recovered during unloading. NiFeGaCo single crystals with cobalt content ranging from 0 to 6% are characterized by a wide SE temperature range, large reversible strain, narrow stress hysteresis, and high cyclic stability of SE (up to 5×10^4 cycles) [2–6].

Recent studies [7,8] have shown that increasing the cobalt content above 10% led to a strain glass transition in these materials, as in numerous other materials that deviate from stoichiometric compositions and contain a large number of point defects [9,10]. These defects impede the thermal-induced MT but contribute to strain glass transitions from unfrozen strain glass into frozen strain glass during stress-free cooling [7–9]. Such materials are characterized by the absence of thermal-induced transitions, high strength properties, high resistance of the matrix to forming stress-induced martensite, and narrow stress hysteresis. However, strain glass alloys have been poorly studied, especially NiFeGaCo alloys with a cobalt content $C_{Co} \geq 10\%$. Only a few studies of NiFeGaCo alloys with $C_{Co} \geq 10\%$ present their significant scientific potential [7,8,11,12]. NiFeGaCo alloys with a cobalt content of 10% are of interest because they still undergo thermal-induced MT but are characterized by typical features for strain glass transitions, such as a logarithmic dependence of temperature curves of the elastic modulus E(T) and internal friction $\tan\delta(T)$ on frequency [13].

The design of NiFeGaCo shape memory alloys includes studying their cyclic stability, the degradation mechanisms of SE, and creating methods for increasing their resistance to cyclic tests. For example, applying shape memory alloys as sensors, dampers, and actuators requires 10^4 loading/unloading cycles, whereas their application in thermal valves, clamps, and micropincers requires at least 10^5 cooling/heating cycles with stable properties [14]. As for studies on SE cyclic stability in NiFeGaCo alloys, there are a few works [4,6,7] that did not clarify the microstructural mechanisms of degradation for NiFeGaCo alloys. Moreover, the number of cycles in these studies did not exceed 5×10^4 cycles [6]. Therefore, the conclusions of these studies are insufficient to predict the dependence of the SE parameters on the higher number of cycles. Single crystals of NiFeGaCo alloys are more promising for further investigation of the SE cyclic stability during large numbers of cycles, up to 10^5.

The main aspect of the present study is the high cobalt content of 10 at.%. First, the higher cobalt content can cause higher ductility in the material. As shown in our previous work [11], an increase in the cobalt content of up to 15% resulted in weak cyclic stability and increased fragility. The $Ni_{39}Fe_{19}Ga_{27}Co_{15}$ single crystals were destroyed during the first 30–50 cycles. In contrast, the $Ni_{44}Fe_{19}Ga_{27}Co_{10}$ alloy was not destroyed and was weakly degraded over 500 loading/unloading cycles. However, 500 loading/unloading cycles were insufficient to examine cyclic stability and determine the degradation mechanisms of these materials. Second, the increase in cobalt content from 6 to more than 10% led to a change in the crystal lattice of the secondary phase particles (from γ' to ω) [7,11]. Therefore, the study of the effect of ω-phase particles on SE cyclic stability in $Ni_{44}Fe_{19}Ga_{27}Co_{10}$ single crystals is relevant. Choosing a different composition and a different crystal structure of the secondary phase in comparison with the literature allows us to find a commonality of the degradation mechanisms for NiFeGaCo alloys, regardless of the cobalt content and the crystal structure of the secondary phase. Therefore, the aim of this work was to study SE cyclic stability (during 10^5 loading/unloading cycles) in quenched $Ni_{44}Fe_{19}Ga_{27}Co_{10}$ crystals without particles and in aged $Ni_{44}Fe_{19}Ga_{27}Co_{10}$ crystals containing ω-phase particles to clarify the effect of ω-phase particles on SE cyclic stability and to establish the microstructural mechanisms of degradation in heterophase NiFeGaCo alloys.

2. Materials and Methods

These studies were conducted only on single crystals because of the extreme brittleness of NiFeGaCo polycrystals. The high-strength [001]-orientation was chosen for the study. The Bridgman method [15,16] was used to grow the $Ni_{44}Fe_{19}Ga_{27}Co_{10}$ (at.%) single crystals. Single crystal growth was performed on a special installation in a ceramic crucible in an inert He atmosphere. The Bridgman method concludes with the heating of the polycrystal above the melting point and follows with slow cooling. Cooling was started from one end of the crucible, where a single crystal of one crystallographic orientation was grown on the seed along the length of the crucible. The formation of single crystals was confirmed by X-ray diffraction analysis, optical and electron microscopy.

These studies were conducted on single crystals after two thermal treatments. First, samples were annealed at 1448 K for 1 h with subsequent water quenching (quenched crystals). Then, one part of the quenched samples was aged at 773 K for 1 h, followed by quenching (aged crystals). For annealing or aging, the samples in a quartz tube filled with argon were put in a hot area of the furnace (Carbolite STF 15/180, Verder Scientific, Haan, Germany). For quenching, the samples were rapidly dropped into cold water. The annealing temperature was chosen based on our previous works [11,12] to obtain a homogeneous structure. After growth and quenching, the surface of single crystals was studied by SEM using TESCAN VEGA 3 (Tescan Group, Brno-Kohoutovice, Czech Republic). The chemical composition of $Ni_{44}Fe_{19}Ga_{27}Co_{10}$ and the homogeneous structure without admixtures were determined. The aging temperature was chosen for the precipitation of ω-phase particles [7,12]. The temperatures of the forward and reverse MT were determined by using the temperature dependence of the electrical resistance. An electro-mechanical testing frame (Instron VHS 5969, Instron, High Wycombe, UK) and an electro-dynamic testing frame (ElectroPuls E3000, Instron, High Wycombe, UK) were used for cyclic tests under compressive load. The compression samples were parallelepipeds with dimensions of $3 \times 3 \times 6$ mm^3. Cyclic tests were performed at room temperature (295 K) during loading/unloading cycles up to a stress level of 300 MPa. The deformation rate for 1–10^5 cycles was V = $9.0 \times 10^{-2} \cdot s^{-1}$; for cycles above 10^5 the deformation rate was V = $21.0 \times 10^{-2} \cdot s^{-1}$. A Universal Keyence VHX 2000 complex (Keyence CORP, Itasca, IL, USA) with a Linkam TST350 testing frame (Redhill, Surrey RH1 5DZ, UK) for in situ compression was used for metallographic observations. A Hitachi HT-7700 electron microscope (Hitachi, Tokyo, Japan) was used for electron microscopy. For electron microscopy, the samples of 3 mm in diameter were mechanically thinned until 0.1 mm and polished using TenuPol-5 electrolytic thinning installation (Struers S.A.S., Champigny-sur-Marne, France).

3. Results

The microstructures of the quenched and aged single crystals are shown in Figure 1 and correspond to those reported in previous studies [11,12]. High-temperature annealing was conducted at 1448 K, above the order-disorder transition temperature [12], and led to the formation of a mixture of B2 and L2$_1$-structures.

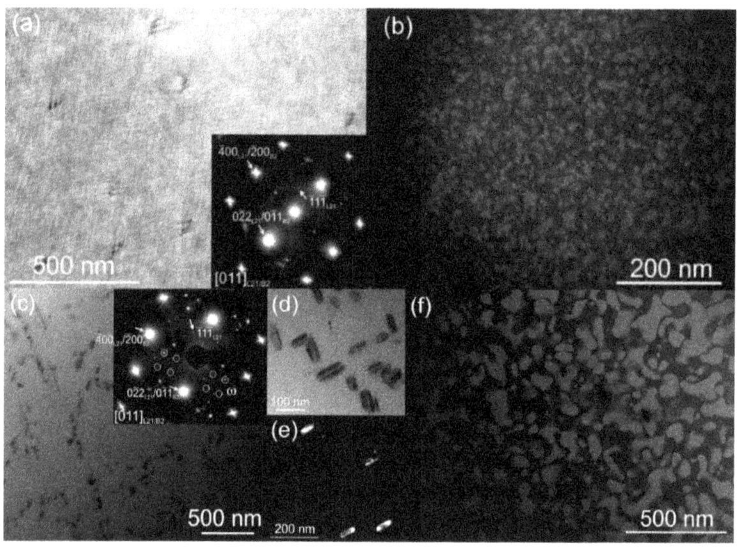

Figure 1. Microstructures of $Ni_{44}Fe_{19}Ga_{27}Co_{10}$ single crystals before cycling: (**a**,**b**) quenched crystals; (**c**–**f**) aged crystals; (**a**,**c**,**d**) bright-field images and SAEDPs with the $[011]_{L21/B2}$ zone axes of the matrix; (**b**,**f**) dark-field images in the $<111>_{L21}$ reflexes marked with arrows in the SAEDPs from (**a**,**c**); dark-field images contain some bright areas, they are L2$_1$-domains; (**e**) dark-field image in the circled reflex on SAEDP from (**c**) showing ω-phase particles.

Both the B2- and L2$_1$-structures have a bcc lattice. The difference between the L2$_1$-lattice and the B2-lattice lies in the ordered arrangement of Fe and Ga atoms at the sites of the sublattice and the lattice parameter [1]. In Figure 1a, the selected area electron diffraction patterns (SAEDPs) for the $[011]_{L2_1}$ and the $[011]_{B2}$ zone axes practically coincide. The lattice parameter of the B2 structure is two times smaller than the lattice parameter of the L2$_1$-structure $a_{L2_1} = 2 \cdot a_{B2}$; therefore, on SAEDP, $r_{hkl}^{L2_1} = 2 \cdot r_{hkl}^{B2}$, hence, the positions of the reflections $[\bar{4}00]_{L2_1}$ and $[\bar{2}00]_{L2_1}$ coincide. In quenched crystals, the mixture of B2 and L2$_1$-structures was visible as nanodomains of 20–50 nm on the dark-field images in the <111> reflections (Figure 1b). The volume fraction of the L2$_1$-structure derived from the dark-field images was estimated to be 40% in the quenched crystals. The fine tweed structure in

Figure 1a is typical for shape memory alloys and was observed in both quenched and aged single crystals.

Aging leads to the precipitation of disordered ω-phase particles with sizes of 70–100 nm and a volume fraction of 4–6% (Figure 1). The reflexes in positions 1/3 and 2/3 $[224]_{L21}$ on SAEDP (Figure 1c) confirm the ω-structure. The reflex in position 1/3 $[224]_{L21}$ corresponds to $[0001]_{\omega}$. Aging causes an increase in the size of the B2 and $L2_1$ phase domains of up to 100–200 nm (Figure 1b,f). The volume fraction of the $L2_1$-structure increases up to 60–65% in the aged crystals compared to the quenched ones. A more detailed description of the microstructure before testing is presented in [12].

The MT temperatures for the quenched crystals were M_s = 203 K, M_f = 163 K, A_s = 173 K, and A_f = 218 K. For the aged crystals, MT temperatures were M_s = 251 K, M_f = 217 K, A_s = 228 K, and A_f = 262 K. The MT temperatures in the aged single crystals were higher than those in the quenched ones, which is associated with the precipitation of particles rich in iron and cobalt. Particles led to decreases in the content of these elements in the matrix and an increase in the MT temperatures. In [12], more details are presented about the change in the MT temperatures after the precipitation of dispersed particles in these crystals.

Cyclic tests were performed at 295 K during loading/unloading cycles at stresses up to 300 MPa, sufficient for complete stress-induced MT in the quenched and aged single crystals (Figure 2). The σ(ε) curves are characterized by different critical stresses σ_{cr}, which are determined by different MT temperatures in accordance with the Clausius–Clapeyron equation. The higher the temperature M_s, the lower the critical stress σ_{cr} at 295 K. The total strains were close, whereas the reversible strains determined at the σ_{cr} differed in the quenched (3.8%) and aged (4.3%) crystals because of the higher stress for MT in the quenched crystals. A decrease in strain with an increase in stress was observed in all NiFeGaCo alloys and is associated with the difference in the elastic moduli of austenite and martensite, as well as with a change in the austenite lattice parameter before the start of MT [17,18].

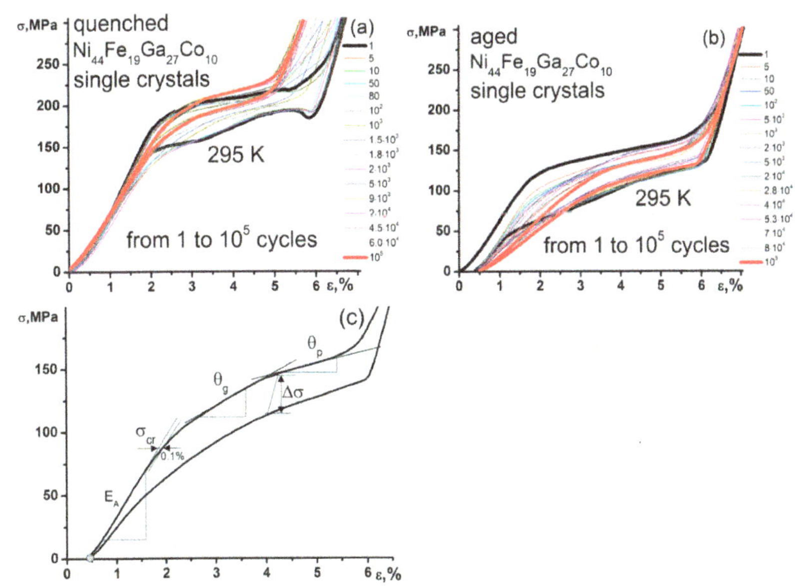

Figure 2. The σ(ε) response: (**a**) 1–10^5 cycles for quenched single crystals; (**b**) 1–10^5 cycles for aged single crystals; the bold black curve is the first cycle; the red curve is the last cycle; (**c**) schematic representation of the σ(ε) curve with measured parameters.

As shown in Figure 2, SE degradation (i.e., a change in the parameters of the $\sigma(\varepsilon)$ curves with an increased number of cycles) was observed in both quenched and aged crystals. The SE degrades in different ways depending on the microstructure. Figure 3 shows the dependence of the SE parameters on the number of cycles: the critical stress, σ_{cr}, the stress hysteresis, $\Delta\sigma$, the effective elastic modulus of austenite, E_A, and the strain hardening coefficient, $\theta = d\sigma/d\varepsilon$ (θ_P and θ_g). These parameters were determined, as shown in Figure 2c.

In Figure 3, the dependences of the SE parameters on the number of cycles in the quenched and aged crystals can be divided into three stages, typical of shape memory alloys [4–6,19–22]. The first stage of initial degradation was observed during the earliest cycles and was characterized by a sharp decrease in the critical stress, σ_{cr}, and the stress hysteresis, $\Delta\sigma$, and a change in the shape of the curves. The second stage of high stability was the longest, and σ_{cr} and $\Delta\sigma$ did not change. The third stage was cyclic fatigue, at which the SE parameters again underwent strong changes. The third stage was observed only in the aged crystals.

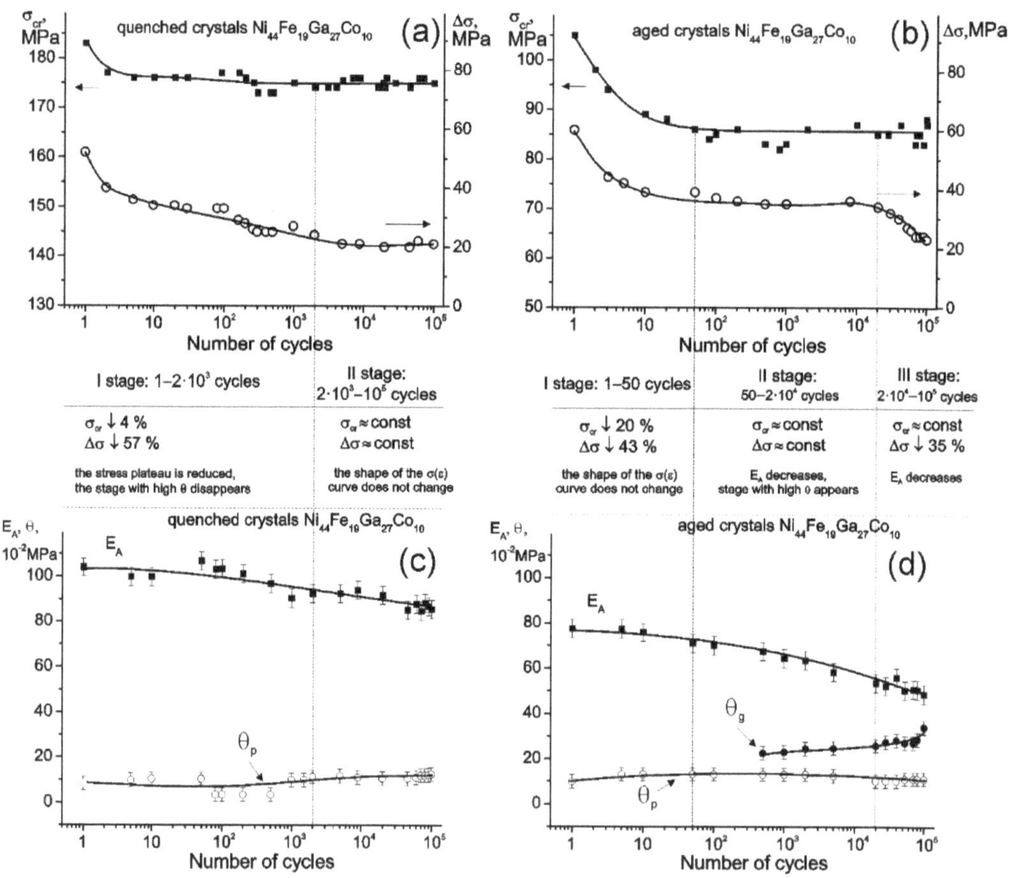

Figure 3. Dependence of critical stress, σ_{cr}, stress hysteresis, $\Delta\sigma$, strain hardening coefficient, $\theta = d\sigma/d\varepsilon$ (θ_p and θ_g), and elastic modulus of austenite, E_A, on the number of cycles (from 1 to 10^5): (**a**,**c**) quenched single crystals; (**b**,**d**) aged single crystals.

In the quenched single crystals, the first stage lasted from 1 to 2×10^3 cycles. In the first stage, σ_{cr} and $\Delta\sigma$ decreased by 4% and 57%, respectively. The $\sigma(\varepsilon)$ curves were characterized by a stress plateau with a low strain hardening coefficient, $\theta_P = d\sigma/d\varepsilon$, close to zero. In this case, the shape of the $\sigma(\varepsilon)$ curve changed greatly. After the stress plateau, the $\sigma(\varepsilon)$ curve showed a segment with high $\theta = d\sigma/d\varepsilon$ and a wide hysteresis $\Delta\sigma$. With an increasing number of cycles, the coefficient $\theta = d\sigma/d\varepsilon$ increased, whereas the strain for a given segment decreased, reducing the hysteresis and the total reversible strain without the appearance of irreversibility. The reasons for the change in the form of the curve require further research. The second stage of high cyclic stability in the quenched single crystals was the longest and lasted from 2×10^3 to 1.2×10^5 cycles. In the second stage, the critical stress, σ_{cr}, the stress hysteresis, $\Delta\sigma$, and the effective elastic modulus of the austenite, E_A, almost did not change. The shape of the curve did not change during the second stage. The third stage was not reached in quenched crystals.

A different pattern of SE degradation was observed in the aged crystals. The first stage of the initial degradation was more than 10 times shorter than that of the quenched crystals and lasted from 1 to 50 cycles. In the first stage, a more noticeable decrease of 20% in the critical stress, σ_{cr}, was observed in the aged crystals, which was five times greater than in the quenched ones. The decrease in the hysteresis, $\Delta\sigma$, by 43% was comparable to the quenched crystals. The form of the $\sigma(\varepsilon)$ curve did not change. In aged single crystals, an irreversible strain of 0.5% was recorded during the first cycles.

The second stage of cyclic stability in the aged crystals was five times shorter than that of the quenched crystals and lasted from 50 to 2×10^4 cycles. The critical stress, σ_{cr}, and the stress hysteresis, $\Delta\sigma$, did not change during the second stage, but there was a significant change in the shape of the $\sigma(\varepsilon)$ curve: a significant decrease in the effective elastic modulus of austenite, E_A, by 25% and the appearance of segments in the $\sigma(\varepsilon)$ curves with different coefficients $\theta = d\sigma/d\varepsilon$. During the first cycles, the forward MT on the $\sigma(\varepsilon)$ curve takes one stage with a stress plateau; after 5×10^2 cycles, the forward MT takes two stages. After reaching the critical stress, a new segment with a high coefficient ($\theta_g = d\sigma/d\varepsilon \approx 20 \times 10^2$ MPa) appeared on the curve. A low-θ_p stress plateau followed the high-θ_g segment. When the number of cycles increased from 5×10^2 to 2×10^4, the coefficient θ_g increased by 25%, while the coefficient $\theta_p = d\sigma/d\varepsilon$ on the stress plateau remained constant.

In aged crystals, a third stage of cyclic fatigue was observed after 2×10^4 cycles. The critical stress changed little at this stage, but the hysteresis was significantly reduced by 35% compared to the second stage. The shape of the curve continued to change. The elastic modulus, E_A, decreased by 11%. The segment of the curve with high-$\theta_g = d\sigma/d\varepsilon$ decreased in the strain by almost two times, while $\theta_g = d\sigma/d\varepsilon$ increased by 35% compared to the second stage of cyclic stability (Figure 3d). A similar change in the SE parameters was also observed in aged $Ni_{49}Fe_{18}Ga_{27}Co_6$ single crystals [6].

4. Discussion

The presence of ω-phase particles was the main reason for the changes in the degradation processes and cyclic stability during the transition from quenched to aged $Ni_{44}Fe_{19}Ga_{27}Co_{10}$ single crystals. First, the particles are quite large, up to 70–100 nm; therefore, they could partially lose coherence with the matrix. In this case, the particles act as stress concentrators and places for dislocation accumulation and can be nucleation sites for non-oriented martensite. Second, these particles led to a decrease in the cobalt content in the matrix, which reduces the brittleness typical of NiFeGaCo alloys with high cobalt content [7–9]. The difference in the resistance to cyclic tests in single crystals with various cobalt contents has been researched in [11]. We have studied the SE in quenched $Ni_{44}Fe_{19}Ga_{27}Co_{10}$ and $Ni_{39}Fe_{19}Ga_{27}Co_{15}$ single crystals and found that high cobalt content results in strong degradation of the SE at the first stage and destruction of the samples within 30–50 cycles [11].

It should be noted that the B2 + $L2_1$ nanodomain structure also has a significant effect on thermal- and stress-induced MTs. In [12], authors considered two microstructures in $Ni_{44}Fe_{19}Ga_{27}Co_{10}$ single crystals: as-grown single crystals, where there is no nanodomain structure and the high-temperature phase is $L2_1$-austenite, and quenched single crystals with a nanodomain (B2 + $L2_1$)-microstructure, as in the present work. It has been shown that for the microstructure containing nanodomains, the stored elastic energy ΔG_{rev} during MT upon cooling/heating is 5–8 times greater than for a single $L2_1$-structure. The stored elastic energy ΔG_{rev} can be estimated from the temperature intervals of the thermal-induced MT [23]. The elastic energy ΔG_{rev} remained high even after aging due to the nanodomain structure (only the size of the domains changed). The high level of stored elastic energy during MT determines the high cyclic stability observed in both the quenched and aged single crystals. As can be seen from the experimental data, the quenched and aged single crystals withstood more than 10^5 cycles without destruction. However, they demonstrated different degradation mechanisms owing to their different microstructural parameters. The ω-phase particles determined the differences in the features of the SE cyclic stability at each stage.

4.1. First Stage of Initial Degradation

In the initial degradation stage in quenched and aged $Ni_{44}Fe_{19}Ga_{27}Co_{10}$ single crystals, the elastic stresses relaxed near the austenite/martensite interface due to the formation of dislocations, residual martensite, and martensite stabilization [4–6,19–22]. In this case, in each subsequent cycle, martensite will be nucleated on the defects remaining after the previous cycle. This facilitates the nucleation of stress martensite and leads to a decrease in critical stress.

In the quenched crystals, the change in critical stress at the first stage of degradation was small, only 4%, whereas, in the aged crystals, the decrease in stress was five times greater. This suggests that the volume fraction of residual martensite and the number of defects were greater in the aged crystals than in the quenched crystals. Indeed, the $\sigma(\varepsilon)$ curves in the first stage of the initial degradation showed an irreversible strain of 0.5% in the aged crystals, in contrast to the quenched crystals without irreversible strain. Notably, the number of cycles at the first stage of degradation in the aged crystals was 10 times less. Thus, more intensive degradation occurred in the aged crystals in fewer cycles than in the quenched crystals.

4.2. Second Stage of High Cyclic Stability

The second stage (Figure 3) was characterized by high stability in the quenched and aged crystals. This stage was five times shorter for the aged single crystals than for the quenched crystals. Although the level of critical stress, σ_{cr}, and the stress hysteresis, $\Delta\sigma$, practically did not change, the shape of the curves changed significantly during the second stage in the aged crystals. There was a decrease in the elastic modulus of austenite E_A; an additional segment appears on the curve with a high coefficient θ_g.

The observed change in the elastic modulus of austenite with an increase in the number of cycles above 5×10^2 in the aged single crystals can be associated with the following processes. First, the microstructure of the aged single crystals changed, and defects and residual martensite appeared. This microstructure differs from defect-free austenite, which was subjected to elastic deformation during the first loading/unloading cycle. Therefore, a mixture of austenite and residual martensite with a high content of stress concentrators (dislocations, residual martensite, and ω-phase particles) is subjected to elastic deformation. Second, stress concentrators are the sites of martensite nucleation. As shown in [24], the stress concentration regions could locally eliminate the nucleation barrier and thus induce equilibrium nanosized embryos of martensite. Consequently, martensite nuclei can appear at a stress below the critical stress, σ_{cr}. This assumption was confirmed by optical observations of the surface of the aged single crystals after 2×10^3 loading/unloading cycles (Figure 4). Without the applied stress, residual martensite was observed in these samples (Figure 4a). As the compressive load increased to small values of 21–63 MPa < σ_{cr}, the residual martensite crystals increased in size, and new stress-induced martensite crystals grew from them.

As shown in Figure 4a, the martensite variants had different orientations. The appearance of non-oriented martensite variants could be caused by the internal stress fields from particles, dislocations, and residual martensite. Following the Clausius–Clapeyron equation, these internal stress fields led to a local increase in the temperature of nucleation of the thermal-induced martensite. For $Ni_{44}Fe_{19}Ga_{27}Co_{10}$ alloys, the coefficient describing the increase in the temperature, M_s^σ, with increasing stress is very low and equals 1.2–1.3 MPa/K [11]. In other words, small internal stresses can significantly shift the temperature of thermal-induced martensite. Then, the martensite observed in Figure 4a may be partly residual and partly thermal-induced, appearing due to the internal stress fields. Indeed, the branched structure of different orientations, as shown in Figure 4a, is characteristic of thermal-induced martensite. If the applied stress increases, this martensite can interact with oriented stress-induced martensite crystals that arise at σ_{cr} and higher. The variant–variant interaction led to an additional segment on the curve with a high coefficient θ_g, which was observed in the aged crystals after 10^3 cycles (Figures 2–4). The additional segment on the curve was related to MT in the part of the sample that has undergone the most plastic deformation, namely, in the part where the greatest number of defects and residual martensite were contained. After the completion of the MT in this region, the transformation occurred in the regions that were less exposed to plastic deformation. This process was accompanied by a stress plateau on the $\sigma(\varepsilon)$ curve. The variant–variant interaction played a more important role in increasing the number of cycles caused by the growth of dislocations and volume fraction of residual martensite. This led to an increase in coefficient θ_g at the second and third stages (Figure 2d).

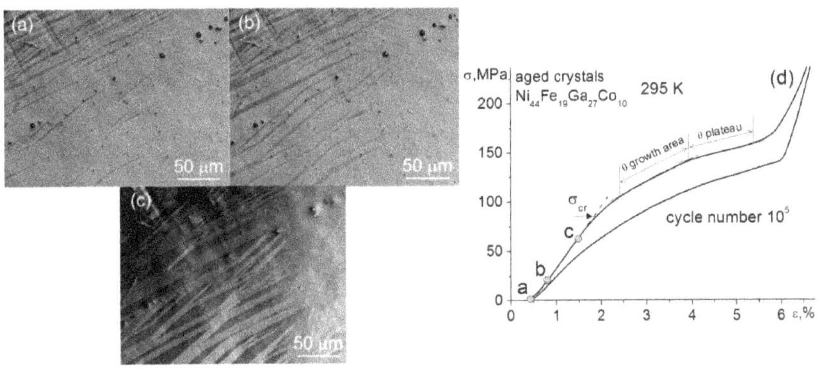

Figure 4. In situ compressive loading for aged $Ni_{44}Fe_{19}Ga_{27}Co_{10}$ single crystals: (**a**–**c**) optical metallography of the surface during in situ compressive loading from 0 to 63 MPa; (**d**) $\sigma(\varepsilon)$ curve at cycle number 2×10^3 at 295 K.

4.3. Third Stage of Cyclic Fatigue and Changes in the Microstructure after Cycling

The quenched single crystals did not reach the third stage of cyclic fatigue within 10^5 cycles. Additional tests showed that the quenched single crystals underwent brittle fracture after 1.2×10^5 cycles without significantly changing the SE parameters. In contrast, after 2×10^4 cycles, the aged single crystals showed the third stage of cyclic fatigue and failed after 1.7×10^5 cycles. In this case, the stress hysteresis, $\Delta\sigma$, the elastic modulus of austenite, E_A, and the coefficient θ_g changed significantly with the number of cycles (Figure 3). These changes are correlated. The decrease in the elastic modulus of austenite, E_A, and the increase in the high-θ_g segment of the curves strongly reduced the low-θ_p stress plateau. The decrease in transformation strain at the low-θ_p stress plateau led to the decrease in stress hysteresis because the smaller volume fraction of the matrix underwent the MT, increasing the number of cycles. A similar stage of cyclic fatigue has also been observed in aged single crystals with a lower cobalt content, $Ni_{49}Fe_{18}Ga_{27}Co_6$ (at.%), after 2×10^4 cycles [6]. The aged $Ni_{49}Fe_{18}Ga_{27}Co_6$ single crystals contained

γ'-phase fine particles smaller than 30 nm and large γ-phase particles of size 5−15 μm, which deformed during testing. The observed dependence of the SE parameters on the number of cycles may be a general property of NiFeGaCo single crystals containing secondary phase particles.

Figures 5 and 6 show the electron microscopy of the quenched and aged single crystals after 10^5 loading/unloading cycles. The degradation of the microstructure in the quenched and aged single crystals proceeded in different ways. First, a higher density of dislocations (Figure 6a) occurred in the aged single crystals than in the quenched ones; the particles served as places for dislocations accumulation (Figure 6b), and residual martensite was observed (Figure 6c). Residual martensite was not found in the quenched crystals using either optical or electron microscopy. It should be mentioned that the difficulties with the pinning and stabilization of martensite are typical for strain glass alloys and were faced frequently in TiNi strain glass alloys. It can also be revealed in quenched $Ni_{44}Fe_{19}Ga_{27}Co_{10}$ single crystals because of high cobalt content and vicinity to the strain glass composition. In contrast, in aged single crystals with second phase particles reducing the cobalt content, martensite is easily pinned by dislocations and particles.

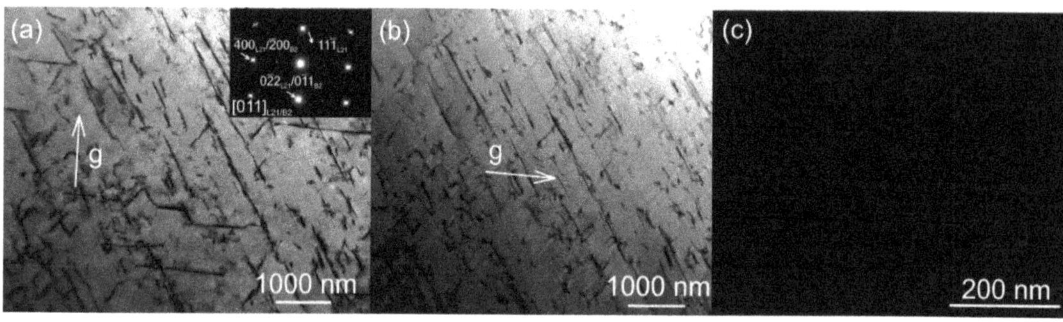

Figure 5. Microstructure of quenched $Ni_{44}Fe_{19}Ga_{27}Co_{10}$ single crystals after cycling: (**a**,**b**) bright-field images and SAEDPs with the $[011]_{L21/B2}$ zone axis of the matrix; (**c**) dark-field image in the <111>$_{L21}$ reflex marked with an arrow in the SAEDP.

Second, no interaction between the nanodomain structure and dislocations was observed in the quenched crystals (Figure 5c), unlike in the aged crystals (Figure 6d). A separation of the nanodomains of the austenitic $L2_1$-structure was observed in the aged single crystals. From experimental studies [25], the slip systems in Ni_2FeGa austenite are likely to be (011)<111> and (011)<001>. In this case, the (011)<111> system is preferable because it is characterized by a lower energy barrier and low activation stress. The slip for this system for the B2 structure would consist of two superpartial dislocations connected by two nearest-neighbor and next-nearest-neighbor antiphase nanodomain boundaries [26]. Refs. [27,28] considered the case where the edge superpartial dislocation pair glided on a slip plane across an antiphase nanodomain boundary perpendicular to the Burgers vector. When the leading superpartial dislocation penetrates the antiphase nanodomain boundary under the influence of the applied stress, a thin band of perfect crystal forms [27,28]. The cases in Figure 6d occurred where the trailing superpartial dislocation had not yet reached the position of the antiphase nanodomain boundary. When the trailing superpartial dislocation met the antiphase nanodomain boundary, further forward motion required the creation of an additional area of the antiphase nanodomain as the dislocation crossed the band of the perfect crystal.

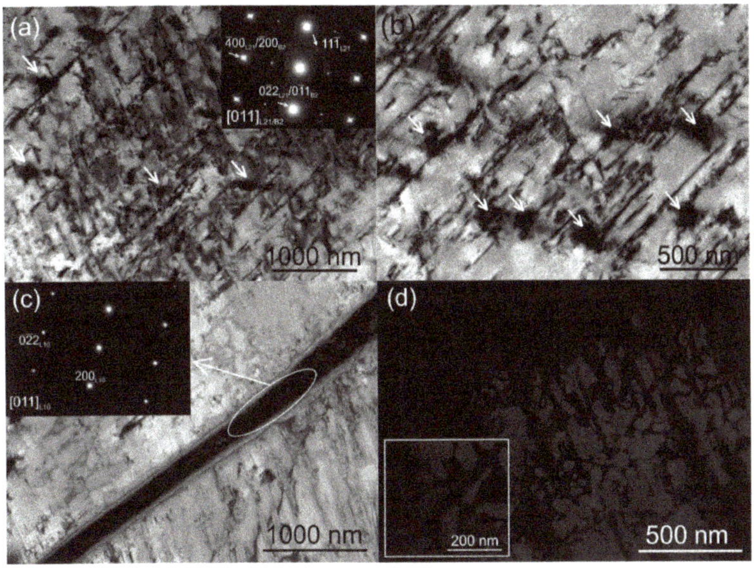

Figure 6. Microstructure of aged $Ni_{44}Fe_{19}Ga_{27}Co_{10}$ single crystals after cycling: (**a**,**b**) bright-field images and SAEDPs with the $[011]_{L21/B2}$ zone axes of the matrix, ω-particles marked with arrows; (**c**) bright-field image with $L1_0$-martensite lamella, $[011]_{L10}$ zone axis of the martensite; (**d**) dark-field image in the $<111>_{L21}$ reflex marked with an arrow in the SAEDP from (**a**), dark-field image showing bright $L2_1$-domains with tracks from the dislocation motion.

It can be concluded from the foregoing that SE degradation occurs differently in quenched and aged single crystals. In quenched crystals, the main degradation mechanism is the formation of a uniform distribution of dislocations (Figure 7). In aged single crystals, the degradation mechanism differs and consists of a non-uniform distribution of dislocations around the particles and the appearance of residual martensite. The ω-phase particles play the main role in aged crystals. These particles serve as additional stress concentrators and reduce the cobalt content, embrittling the matrix. There were a large number of defects and residual martensite fixed around the particles (Figure 7). Therefore, in aged single crystals, the relaxation of elastic stresses during cyclic tests was more active than in quenched crystals, as evidenced by the more significant change in the SE parameters over a smaller number of cycles (Figures 2 and 3). In contrast, the quenched crystals contain smaller stress concentrators that contribute to the relaxation of elastic stresses. As a result, locally high stresses that did not relax led to sample destruction without the third stage of degradation.

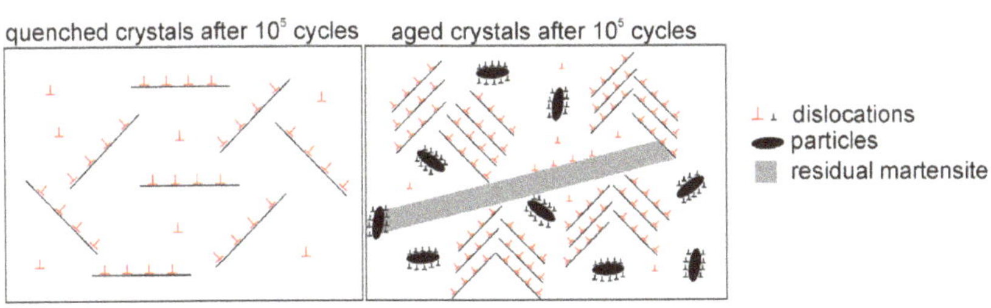

Figure 7. Scheme of microstructures of aged and quenched $Ni_{44}Fe_{19}Ga_{27}Co_{10}$ single crystals after cycling.

5. Conclusions

It was experimentally established that quenched and aged $Ni_{44}Fe_{19}Ga_{27}Co_{10}$ single crystals (at.%) withstood 10^5 loading/unloading cycles of SE at 295 K without breaking, which makes it promising for applications.

For the quenched crystals (quenched after annealing at 1448 K for 1 h), the degradation process of SE consisted of two stages: the first stage of the initial degradation with a significant decrease in the SE parameters and the second stage of high cyclic stability and small parameters change. The main degradation mechanism was the formation of a uniform distribution of dislocations.

Aging at 773 K for 1 h caused the precipitation of particles of the ω-phase with a size of 70–100 nm, which led to the change of both the degradation process of SE and the microstructural degradation mechanisms. The degradation process was more intensive and occurred in smaller number of cycles with an additional cyclic fatigue stage, where SE parameters decreased sharply. The microstructural degradation mechanisms consist not only of a non-uniform distribution of dislocations but also include accumulation of the dislocations around the particles, which can contribute to pinning the residual martensite.

Author Contributions: Conceptualization, E.T., E.P. and Y.I.C.; methodology, Y.I.C.; validation, A.T. and M.Z.; formal analysis, E.P. and E.T.; investigation, M.Z.; resources, A.T. and M.Z.; writing—original draft preparation, E.T.; writing—review and editing, E.P., E.T. and Y.I.C.; visualization, E.T.; supervision, E.P. and Y.I.C.; project administration, E.T. All authors have read and agreed to the published version of the manuscript.

Funding: This research was funded by the Russian Science Foundation (grant No. 21-19-00287).

Data Availability Statement: The data used in this article are presented in the manuscript.

Acknowledgments: The electron microscopy studies were carried out on the equipment of the Krasnoyarsk Regional Center for Collective Use SB RAS.

Conflicts of Interest: The authors declare no conflict of interest.

References

1. Hamilton, R.; Sehitoglu, H.; Efstathiou, C.; Maier, H. Inter-martensitic transitions in Ni–Fe–Ga single crystals. *Acta Mater.* **2007**, *55*, 4867. [CrossRef]
2. Masdeu, F.; Pons, J.; Torrens-Serra, J.; Chumlyakov, Y.; Cesari, E. Superelastic behavior and elastocaloric effect in a $Ni_{51.5}Fe_{21.5}Ga_{27.0}$ ferromagnetic shape memory single crystal under compression. *Mater. Sci. Eng. A* **2022**, *833*, 142362. [CrossRef]
3. Masdeu, F.; Pons, J.; Chumlyakov, Y.; Cesari, E. Two-way shape memory effect in $Ni_{49}Fe_{18}Ga_{27}Co_6$ ferromagnetic shape memory single crystals. *Mater. Sci. Eng. A* **2021**, *805*, 140543. [CrossRef]
4. Efstathiou, C.; Sehitoglu, H.; Kurath, P.; Foletti, S.; Davoli, P. Fatigue response of NiFeGa single crystals. *Scr. Mater.* **2007**, *57*, 409–412. [CrossRef]
5. Imran, M.; Zhang, X. Recent developments on the cyclic stability in elastocaloric materials. *Mater. Des.* **2020**, *195*, 109030. [CrossRef]
6. Larchenkova, N.; Panchenko, E.; Timofeeva, E.; Tagiltsev, A.; Chumlyakov, Y. Cyclic stability of superelasticity in [001]-oriented stress-free and stress-assisted aged $Ni_{49}Fe_{18}Ga_{27}Co_6$ single crystals. *AIP Conf. Proc.* **2008**, *2051*, 020166.
7. Chen, H.; Wang, Y.; Nie, Z.; Li, R.; Cong, D.; Liu, W.; Ye, F.; Liu, Y.; Cao, P.; Tian, F.; et al. Unprecedented non-hysteretic superelasticity of [001]-oriented NiCoFeGa single crystals. *Nat. Mater.* **2020**, *19*, 712–718. [CrossRef]
8. Zhang, Q.; Zhai, Z.; Nie, Z.; Harjo, S.; Cong, D.; Wang, M.; Lie, J.; Wang, Y. An in situ neutron diffraction study of anomalous superelasticity in a strain glass $Ni_{43}Fe_{18}Ga_{27}Co_{12}$ alloy. *J. Appl. Crystallogr.* **2015**, *48*, 1183. [CrossRef]
9. Sun, X.; Cong, D.; Ren, Y.; Liss, K.-D.; Brown, D.E.; Ma, Z.; Hao, S.; Xia, W.; Chen, Z.; Ma, L.; et al. Magnetic-field-induced strain-glass-to-martensite transition in a Fe-Mn-Ga alloy. *Acta Mater.* **2020**, *183*, 11–23. [CrossRef]
10. Hao, C.; Wang, Y.; Wu, X.; Guo, Y.; Liu, C.; He, L.; Adil, M.; Yang, S.; Song, X. High performance damping behavior of Ni-Fe-Ga alloys within the martensite/strain-glass phase boundary. *J. Alloys Compd.* **2022**, *898*, 162954. [CrossRef]
11. Timofeeva, E.; Panchenko, E.; Eftifeeva, A.; Tagiltsev, A.; Surikov, N.; Tokhmetova, A.; Yanushonite, E.; Zherdeva, M.; Karaman, I.; Chumlyakov, Y. Cyclic stability of superelasticity in [001]-oriented quenched $Ni_{44}Fe_{19}Ga_{27}Co_{10}$ and $Ni_{39}Fe_{19}Ga_{27}Co_{15}$ single crystals. *Acta Metall. Sin.* **2023**, *36*, 650–660. [CrossRef]
12. Timofeeva, E.; Panchenko, E.; Zherdeva, M.; Tokhmetova, A.; Surikov, N.; Chumlyakov, Y.; Karaman, I. The effect of thermal treatment on microstructure and thermal-induced martensitic transformations in $Ni_{44}Fe_{19}Ga_{27}Co_{10}$ single crystals. *Metals* **2022**, *12*, 1960. [CrossRef]

13. Timofeeva, E.E.; Panchenko, E.Y.; Zherdeva, M.V.; Chumlyakov, Y.I.; Karaman, I. Thermal- and stress-induced martensitic transformations in [001]-oriented $Ni_{44}Fe_{19}Ga_{27}Co_{10}$ single crystals. *Mater. Lett.* **2022**, *310*, 131477. [CrossRef]
14. Hornbogen, E. Review Thermo-mechanical fatigue of shape memory alloys. *J. Mater. Sci.* **2004**, *39*, 385–399. [CrossRef]
15. Dutta, P.S. Bulk growth of crystals of III–V compound semiconductors. *Semicond. Sci. Technol.* **2011**, *3*, 36–80.
16. Venkataraman, R. Semiconductor detectors. In *Handbook of Radioactivity Analysis*, 4th ed.; L'Annunziata, M.F., Ed.; Academic Press: Cambridge, MA, USA, 2020; Volume 1, pp. 409–491.
17. Ding, X.D.; Suzuki, T.; Suna, J.; Ren, X.; Otsuka, K. Study on elastic constant softening in stress-induced martensitic transformation by molecular dynamics simulation. *Mater. Sci. Eng. A* **2006**, *438–440*, 113–117. [CrossRef]
18. Lauhoff, C.; Reul, A.; Langenkämper, D.; Krooß, P.; Somsen, C.; Gutmann, M.; Pedersen, B.; Kireeva, I.; Chumlyakov, Y.; Eggeler, G.; et al. Effects of aging on the stress-induced martensitic transformation and cyclic superelastic properties in Co-Ni-Ga shape memory alloy single crystals under compression. *Acta Mater.* **2022**, *226*, 117623. [CrossRef]
19. Li, X.; Chen, H.; Guo, W.; Guan, Y.; Wang, Z.; Zeng, Q.; Wang, X. Improved superelastic stability of NiTi shape memory alloys through surface nano-crystallization followed by low temperature aging treatment. *Intermetallics* **2021**, *131*, 107114. [CrossRef]
20. Gall, K.; Maier, H. Cyclic deformation mechanisms in precipitated NiTi shape memory alloys. *Acta Mater.* **2002**, *50*, 4643–4657. [CrossRef]
21. Krooß, P.; Kadletz, P.; Somsen, C.; Gutmann, M.; Chumlyakov, Y.; Schmahl, W.; Maier, H.; Niendorf, T. Cyclic degradation of $Co_{49}Ni_{21}Ga_{30}$ high-temperature shape memory alloy: On the roles of dislocation activity and chemical order. *Shape Mem. Superelast.* **2016**, *2*, 37–49. [CrossRef]
22. Kann, Q.; Yu, C.; Kang, G.; Li, J.; Yan, W. Experimental observations on rate-dependent cyclic deformation of superelastic NiTi shape memory alloy. *Mech. Mater.* **2016**, *97*, 48–58. [CrossRef]
23. Beke, D.; Daróczi, L.; Samy, N.; Tóth, L.; Bolgár, M. On the thermodynamic analysis of martensite stabilization treatments. *Acta Mater.* **2020**, *200*, 490–501. [CrossRef]
24. Rao, W.; Xu, Y.; Hu, C.; Khachaturyan, A. Magnetoelastic equilibrium and super-magnetostriction in highly defected pre-transitional materials. *Acta Mater.* **2020**, *188*, 539–550. [CrossRef]
25. Sehitoglu, H.; Wang, J.; Maier, H. Transformation and slip behavior of Ni_2FeGa. *Int. J. Plast.* **2012**, *39*, 61–74. [CrossRef]
26. Chowdhury, P.; Sehitoglu, H. A revisit to atomistic rationale for slip in shape memory alloys. *Prog. Mater. Sci.* **2017**, *85*, 1–42. [CrossRef]
27. Koizumi, Y.; Minamino, Y.; Nakano, T.; Umakoshi, Y. Effects of antiphase domains on dislocation motion in Ti_3Al single crystals deformed by prism slip. *Philos. Mag.* **2008**, *88*, 465–468. [CrossRef]
28. Rong, T.; Aindow, M.; Jones, I. The interaction between extended dislocations and antiphase domain boundaries—I: Superpartial separation and the yield stress. *Intermetallics* **2001**, *9*, 499–506. [CrossRef]

metals

MDPI

Article

Magnetostriction of Heusler Ferromagnetic Alloy, $Ni_2MnGa_{0.88}Cu_{0.12}$, around Martensitic Transition Temperature

Takuo Sakon [1,*], Koki Morikawa [1], Yasuo Narumi [2], Masayuki Hagiwara [2], Takeshi Kanomata [3], Hiroyuki Nojiri [4] and Yoshiya Adachi [5]

1 Mechanics and Robotics Course, Faculty of Advanced Science and Technology, Ryukoku University, Otsu 520-2194, Shiga, Japan
2 Center for Advanced High Magnetic Fields, Graduated School of Science, Osaka University, Toyonaka 560-0043, Osaka, Japan; narumi@ahmf.sci.osaka-u.ac.jp (Y.N.); hagiwara@ahmf.sci.osaka-u.ac.jp (M.H.)
3 Research Institute for Engineering and Technology, Tohoku Gakuin University, Tagajo 985-8537, Miyagi, Japan; c1924007@mail.tohoku-gakuin.ac.jp
4 Institute for Materials Research, Tohoku University, Sendai 980-8577, Miyagi, Japan; hiroyuki.nojiri.e8@tohoku.ac.jp
5 Graduate School of Science and Engineering, Yamagata University, Yonezawa 992-8510, Yamagata, Japan; adachy@yz.yamagata-u.ac.jp
* Correspondence: sakon@rins.ryukoku.ac.jp

Abstract: In this study, magnetostriction measurements were performed on the ferromagnetic Heusler alloy, $Ni_2MnGa_{0.88}Cu_{0.12}$, which is characterized by the occurrence of the martensitic phase and ferromagnetic transitions at the same temperature. In the austenite and martensite phases, the alloy crystallizes in the $L2_1$ and $D0_{22}$-like crystal structure, respectively. As the crystal structure changes at the martensitic transition temperature (T_M), a large magnetostriction due to the martensitic and ferromagnetic transitions induced by magnetic fields is expected to occur. First, magnetization (*M-H*) measurements are performed, and metamagnetic transitions are observed in the magnetic field of μ_0H = 4 T at 344 K. This result shows that the phase transition was induced by the magnetic field under a constant temperature. Forced magnetostriction measurements ($\Delta L/L$) are then performed under a constant temperature and atmospheric pressure (P = 0.1 MPa). Magnetostriction up to 1300 ppm is observed around T_M. The magnetization results and magnetostriction measurements showed the occurrence of the magnetic-field-induced strain from the paramagnetic austenite phase to the ferromagnetic martensite phase. As a reference sample, we measure the magnetostriction of the Ni_2MnGa-type ($Ni_{50}Mn_{30}Ga_{20}$) alloy, which causes the martensite phase transition at T_M = 315 K. The measurement of magnetostriction at room temperature (298 K) showed a magnetostriction of 3300 ppm. The magnetostriction of $Ni_2MnGa_{0.88}Cu_{0.12}$ is observed to be one-third that of $Ni_{50}Mn_{30}Ga_{20}$ but larger than that of Terfenol-D (800 ppm), which is renowned as the giant magnetostriction alloy.

Keywords: Heusler alloy; Ni_2MnGa; ferromagnetism; shape memory alloy; magnetostriction; martensite

1. Introduction

Ferromagnetic Ni_2MnGa-type Heusler alloys are known as highly functional materials in applications such as smart actuators and robotics, which utilizes the magnetic field-induced strain (MFIS) [1–30], and magnetic refrigeration, which utilizes the magnetocaloric effect [31–36].

In our previous study, we performed thermal expansion, permeability, and magnetization measurements on $Ni_2MnGa_{0.88}Cu_{0.12}$ [37], which is commonly characterized by the occurrence of martensitic and ferromagnetic transitions at the same temperature. Its permeability abruptly changes and shows a clear peak around the martensitic transition temperature, T_M. The temperature dependence of the magnetization also shows a clear decrease around T_M. The value of T_M obtained through linear expansion was 337 K. Furthermore, T_M and T_R (reverse martensitic transition temperature from martensite phase to austenite phase) increase gradually with increasing magnetic fields.

T_M of the ferromagnetic Heusler alloy, $Ni_2MnGa_{0.88}Cu_{0.12}$, in the magnetic field is considered to shift according to the magnetic fields, and it is proportional to the difference between the magnetization of austenite and martensite phases. The shift of T_M was estimated as $dT_M/d(\mu_0H)$ = 1.3 K/T. Khovailo et al. [38–40] discussed the correlation between the shifts of T_M of $Ni_{2+x}Mn_{1-x}Ga$ ($0 \leq x \leq 0.19$) using theoretical calculations. The experimental values of this shift corresponded well with the theoretical calculation results. In general, in a magnetic field, the Gibbs free energy is lowered by the Zeeman energy $-\Delta MB$, which enhances the motive force of the martensitic transition. Thus, T_M of $Ni_2MnGa_{0.88}Cu_{0.12}$ is considered to have shifted in accordance with the magnetic fields because high magnetic fields are favorable to the ferromagnetic martensitic phases.

The $dT_M/d(\mu_0H)$ of $Ni_2MnGa_{0.88}Cu_{0.12}$ is larger than that of other Ni_2MnGa-type Heusler alloys [18,38,40,41], indicating that the magnetic fields considerably affect the martensitic transition of $Ni_2MnGa_{0.88}Cu_{0.12}$ compared to that of other alloys.

Mendonça et al. investigated the magnetizations and magnetostrictions of off-stoichiometric Ni_2MnGa-type polycrystalline $Ni_2Mn_{0.7}Cu_{0.3}Ga_{0.84}Al_{0.16}$ alloy [19]. At a temperature slightly higher than T_M, a magnetostriction, $\Delta L/L$, caused by the magneto-structural transition demonstrating a significant deformation of 26,000 ppm (2.6%) with metamagnetic transition from the low-magnetic field austenite to the high-magnetic field martensite phase was observed. Martensitic and ferromagnetic transitions occur at the same temperature of T_M = 293 K; this property is similar to that of $Ni_2MnGa_{0.88}Cu_{0.12}$ [37]. The shift of T_M was estimated as $dT_M/d(\mu_0H)$ = 1 K/T, which is of the same order as that of $Ni_2MnGa_{0.88}Cu_{0.12}$. The temperature dependence of the magnetization *M-T* curve of $Ni_2Mn_{0.7}Cu_{0.3}Ga_{0.84}Al_{0.16}$ indicated the presence of a metamagnetic-like large hysteresis at 298 K. This, in turn, indicates that at a constant temperature, re-entrant transitions occur between the paramagnetic austenite phase and ferromagnetic martensite phase. In this study, the *M-T* curve of $Ni_2MnGa_{0.88}Cu_{0.12}$ indicated the metamagnetic-like large hysteresis at 343 and 345 K, which are slightly higher than T_M. Moreover, the metamagnetic behavior suggests that $Ni_2MnGa_{0.88}Cu_{0.12}$ can generate a large magnetostriction around T_M.

In this study, we analyzed the magnetostriction of $Ni_2MnGa_{0.88}Cu_{0.12}$ and compared its characteristics with those of other MFIS alloys.

2. Materials and Methods

2.1. Materials

The polycrystalline $Ni_2MnGa_{0.88}Cu_{0.12}$ sample was grown by repeated arc melting of the constituent elements, namely 4N Ni, 4N Mn, 4N Cu, and 6N Ga, in an Ar atmosphere. All reaction products were packed in evacuated silica tubes at 1123 K for 72 h and then at 873 K for 24 h before being quenched into water [42,43]. The samples were characterized using X-ray powder diffraction measurements based on Cu-Kα radiation. At room temperature, the crystalline structure was a tetragonal $D0_{22}$-like martensite. In addition, the results of the X-ray diffraction indicated that the tetragonal phase with a $D0_{22}$-like crystal structure coexists with the monoclinic phase of the monoclinic *14M* structure (space group: *P2/m*) at room temperature. The lattice parameters of $D0_{22}$ are a_{D022} = 0.38920 nm and c_{D022} = 0.65105 nm. The details of the preparation and characterization of the sample can be found in references [42,43]. As a result of having observed a sample by an optical microscope at 298 K, there are some grains as large as 0.1 mm, which size was smaller than that of polycrystalline $Ni_2Mn_{0.7}Cu_{0.3}Ga_{0.84}Al_{0.16}$ alloy (grain size 0.5 mm) [19].

A Ni_2MnGa-type single-crystalline alloy was purchased from Adaptamat Co., Ltd. (Helsinki, Finland). The materials were analyzed using EDS (6010LA, JEOL Co., Ltd., Akishima, Japan) at the Faculty of Advanced Science and Technology, Ryukoku University. The materials included 29.79 at.% Mn, 50.03 at.% Ni, and 20.18 at.% Ga. Therefore, the alloy was transcribed into $Ni_{50}Mn_{30}Ga_{20}$. An alloy with similar lattice constants and crystal structure as this sample has been presented in [5]. At room temperature, the crystal structure was a 7-layered *7M* martensite (which corresponds to the *14M* structure and P2/m space group) with lattice constants of a = 0.4219 nm, b = 0.5600 nm, and c = 2.0977 nm. At 324 K, the crystal structure was transformed into cubic-$L2_1$ austenite (space group Fm3 m) with a lattice constant of a = 0.5837 nm.

2.2. Methods for Experimental Measurements

Magnetization measurements were performed using the SQUID magnetometer (MPMS-XL7, Quantum Design Co., Ltd., San Diego, CA, USA) at the Center for Advanced High Magnetic Fields, Osaka University. The sample size of $Ni_2MnGa_{0.88}Cu_{0.12}$ for the magnetization measurement was 1.4 × 1.4 × 1.4 mm. The magnetization process *M-H* measurements were performed at temperatures ranging from 360 K to 330 K. Each measurement point took 30 s. Temperature dependence of the magnetization *M-T* measurements were performed by warming the sample from 310 K to 400 K, followed by cooling from 400 K to 310 K. The temperature sweep ratio was 0.016 K/s. Each measurement point took 60 s.

Magnetostriction measurements were carried out using strain gauges KFR-02-120-C1-16 (Kyowa Dengyo Co., Ltd., Chofu, Japan) under atmospheric pressure and without applying any pre-stress. The base size of this strain gauge is 2.5 mm (parallel to the grid) × 2.2 mm. The electrical resistivity of the strain gauges was measured using the four-probe method and was measured by means of 2000/J digital voltmeter (Keithley Co., Ltd., Cleveland, OH, USA). This digital voltmeter was connected to a CF-FX1 Laptop PC (Panasonic Co., Ltd., Kadoma, Japan) via a GPIB cable, and measured by means of Basic 98 program. The sample size of $Ni_2MnGa_{0.88}Cu_{0.12}$ is 3.5 × 3.0 × 3.0 mm. The magnetostrictions of $Ni_2MnGa_{0.88}Cu_{0.12}$ were measured parallel to the applied magnetic field $(\Delta L/L)//H$. The sweep speed of the magnetic field during the magnetostriction measurement was 0.0125 T/s. Each magnetostriction measurement at a specific temperature was measured after heating the sample to 370 K, which is higher than T_R, and then maintaining a constant temperature. In order to estimate the error of the strain gauge, the strain gauge was fixed on Cu plate, and the magnetoresistance of the strain gauge was measured in a magnetic field up to 10 T. The error was estimated to be ±1 ppm at 5 T and ±3 ppm at 10 T.

The forced magnetostriction measurements for $Ni_2MnGa_{0.88}Cu_{0.12}$ were performed using a helium-free superconducting magnet up to 6 T at the Center for Advanced High Magnetic Field Science, Osaka University, and up to 9.8 T at the High Field Laboratory for Superconducting Materials, Institute for Materials Research, Tohoku University.

The forced magnetostriction measurement for $Ni_{50}Mn_{30}Ga_{20}$ was performed using a four-prove strain gauge method up to 1.5 T with a water-cooled electromagnet (Tamagawa Seisakusho Co., Ltd., Sendai, Japan) at the Faculty of Advanced Science and Technology, Ryukoku University. The sample size was 3.5 × 2.5 × 1.0 mm, and the longitudinal direction of the sample was parallel to the c-axis. The magnetostriction was measured along the c-axis in a magnetic field applied parallel to the c-axis, $(\Delta L/L)//H//c$.

3. Results and Discussions

3.1. Temperature and Magnetic Field Dependency of the Magnetization

Figure 1 shows the temperature dependence of the magnetization for $Ni_2MnGa_{0.88}Cu_{0.12}$ in a magnetic field of μ_0H = 0.1 T. The low- and high-temperature phases below and above 340 K indicate the ferromagnetic martensite and paramagnetic austenite phases, respectively. This is consistent with the results of previous studies [37,42,43]. The reason of the dip of the magnetization *M-T* curve around 325 K in cooling process, and the difference of the *M-T* curve in cooling and heating processes shown in Figure 1, can be attributed to the alignment of magnetic moments of Mn and Ni atoms parallel to the magnetic field just below T_M and the rearrangement of the *14M* and/or DO_{22} tetragonal lattices by the magnetic moments.

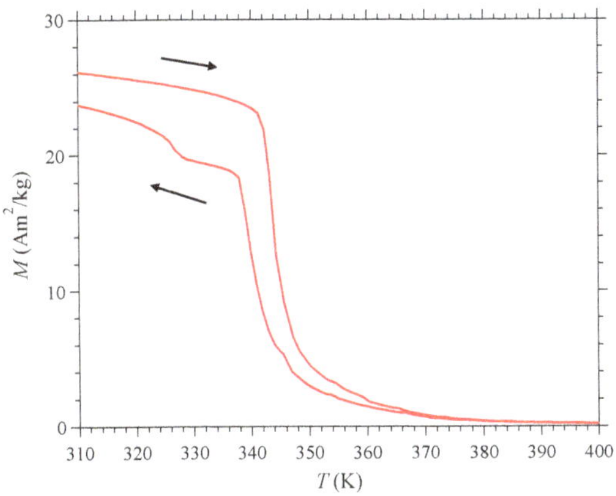

Figure 1. Temperature dependence of the magnetization, *M* vs. *T*, for $Ni_2MnGa_{0.88}Cu_{0.12}$ in a magnetic field of 0.1 T. The arrows show the process of the temperature.

The values of *dM/dT* shown in Figure 2 indicate magnetization with respect to temperature. The peak temperature of *dM/dT* corresponds to the transition temperature, indicating the martensitic transition temperature of T_M = 338 K (austenite phase to martensite phase). The reverse martensitic transition temperature is almost the same as T_M, with a value of 337 K, which was obtained through linear expansion in our previous study [37]. Based on the heating process of *dM/dT*, T_R = 343 K (martensite phase to austenite phase). From the *dM/dT* results shown in Figure 2, the martensitic transition start temperature T_{Ms}, the martensitic transition finish temperature T_{Mf}, the reverse martensitic transition start temperature T_{Rs}, and the reverse martensitic transition finish temperature T_{Rf}, was obtained as T_{Ms} = 349 K, T_{Mf} = 335 K, T_{Rs} = 340 K, T_{Rf} = 353 K. The martensitic temperature T_M as 338 K was obtained from the peak temperature of *dM/dT* during the cooling process. The reverse martensitic transition temperature T_R was obtained from the peak temperature of *dM/dT* in the heating process.

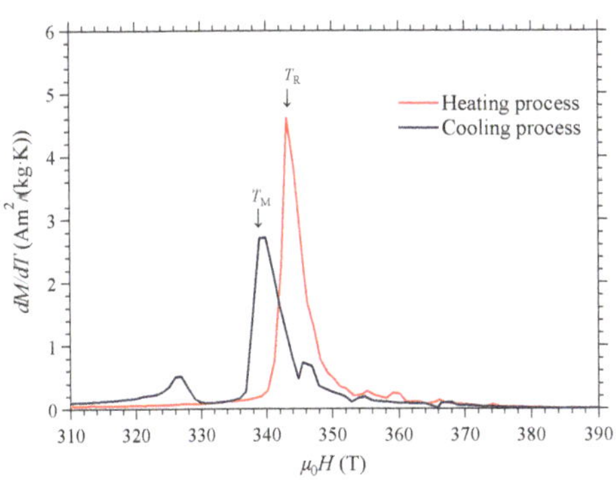

Figure 2. Differential magnetization at temperature, *dM/dT* vs. *T*, for $Ni_2MnGa_{0.88}Cu_{0.12}$ in a magnetic field of 0.1 T. T_M and T_R indicate the martensitic transition temperature and the reverse martensitic transition temperature.

In Figure 3, the magnetic field dependence of the magnetization for $Ni_2MnGa_{0.88}Cu_{0.12}$ at a constant temperature is shown. It can be observed that the martensite phase has larger magnetization than that of the austenite phase. Metamagnetic like hysteresis was observed around 340 K. This indicates that, at the constant temperature, transition from low-magnetic field paramagnetic austenite phase to high-magnetic field ferromagnetic martensite phase occurred. This is similar to the re-entrant transition between paramagnetic austenite phase and ferromagnetic martensite phase as shown in $Ni_2Mn_{0.7}Cu_{0.3}Ga_{0.84}Al_{0.16}$ polycrystalline alloy [19].

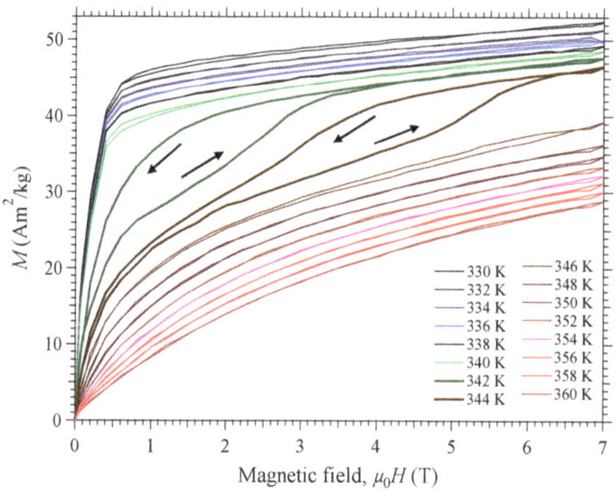

Figure 3. Magnetization process, *M* vs. *H*, for $Ni_2MnGa_{0.88}Cu_{0.12}$ from 360 to 330 K with 2 K steps. The arrows show the process of the magnetic field.

The magnetostriction was measured to investigate the relation between the structural phase transition and magnetic property.

3.2. Forced Magnetostriction

The forced magnetostriction measurements for $Ni_2MnGa_{0.88}Cu_{0.12}$ were performed from a higher temperature to a lower temperature compared to the martensite transition temperature.

Figure 4 shows the magnetic field dependence of the forced magnetostriction, $\Delta L/L$, up to 6 T. First, we measured it at 352 K, which is higher than T_M and then at a constant temperature, while gradually cooling the sample. For the y-axis ($\Delta L/L$), an offset is applied.

The magnetostriction increases when the sample is cooled from 352 K, and reaches its maximum at 1300 ppm near 343 K. Furthermore, the magnetostriction decreases when the sample is cooled. These results indicate that large magnetostriction was observed at a slightly higher temperature than the martensite transition temperature, T_M = 338 K. Furthermore, large hysteresis was observed at 345 and 343 K, which could be equivalent to the hysteresis of magnetization; in addition, re-entrant transition occurred between the paramagnetic austenite and ferromagnetic martensite phases.

As for Ni_2MnGa alloys, the rearrangement of the martensite variants is responsible for a large part of the magnetic field-induced strain. The large plastic strain observed between 341 K and 345 K indicates that this effect plays a significant and important role. In Figure 4, the magnetostriction curve at 352 K and 350 K, in the fully austenite phase, which has a temperature lower than T_{Ms} = 349 K, indicates a small hysteresis. At 348 K, the hysteresis of the magnetostriction curve is larger than that at 350 K. This indicates that the temperature was slightly lower than T_{Ms}, and thus the martensitic transition took place. At 345 K, the largest hysteresis was found. At this temperature, we expect that the austenite parent phase at low magnetic fields would change into the martensite phase. The result of thermal strain in the cooling process indicates a contraction of 1300 ppm (0.13%) with a martensitic transition around T_M [37]. The shrinkage value of the magnetostriction curve is comparable to the contraction value of the thermal strain result observed in our former study. For the isothermal magnetization process at 344 K, as shown in Figure 3, the *M-H* curve for decreasing magnetic field returns to the initial phase before the field reaches zero, suggesting a considerable recovery of the austenite phase. Therefore, a large magnetic field-induced strain with a large hysteresis and significant austenite recovery at 345 K was observed. These results are consistent with those of $Ni_{2.15}Mn_{0.70}Cr_{0.15}Ga$ polycrystal as shown in reference [20]. From the magnetization and the magnetostriction results, the magnetic driving force is expected to be stronger than the potential barrier for the martensitic transition from the austenite phase to the martensite phase.

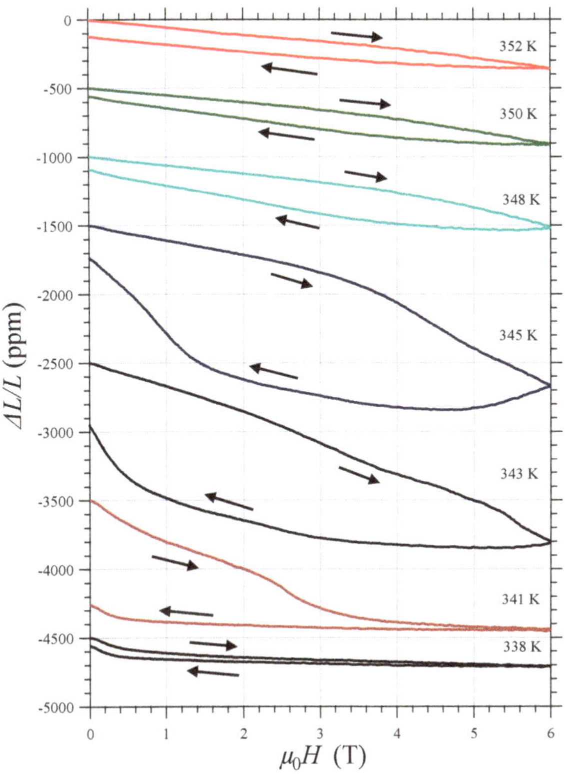

Figure 4. Magnetic field dependence of forced linear magnetostriction, $\Delta L/L$ vs. H, for $Ni_2MnGa_{0.88}Cu_{0.12}$. The arrows show the process of the magnetic field.

At 341 K, which is higher than T_{Mf} = 335 K, there are still some austenite regions in zero magnetic field. Above 4 T, the strain value was almost constant, which indicates that fully martensitic state was realized, and the variants were fully aligned in a high magnetic field. As the magnetic field-induced martensitic transition occurred at high magnetic fields, it is conceived that the recovery strain became small as it is difficult to exceed the potential barrier in low magnetic fields.

At 338 K, a nearly fully martensite state was realized even at zero magnetic field, resulting in small magnetostriction.

Figure 5 shows the magnetic field dependence of the forced magnetostriction, $\Delta L/L$, up to 9.8 T. At 345 K, the re-entrant transition is clearly identified. At 347 K, the transition shrank almost linearly with increasing magnetic fields, demonstrating hysteresis. At 345 K, the sample also shrank almost linearly with increasing magnetic fields. The shrinkage decreased beyond 8 T. When the magnetic field decreased from 9.8 T, the length of the sample remained almost the same, and below 6 T, the sample gradually lengthened. At 1.5 T, a soft kink was observed in $\Delta L/L$. The result of magnetostriction at 345 K indicates that the magnetic-field-induced transition from the paramagnetic austenite to ferromagnetic martensite phase occurred in increasing magnetic field process.

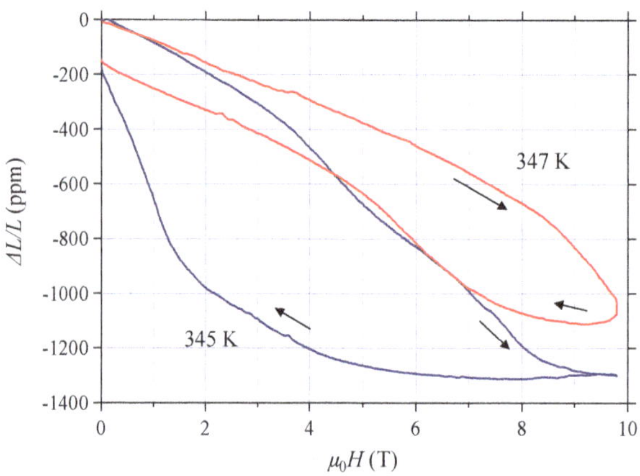

Figure 5. Magnetic field dependence of forced linear magnetostriction, $\Delta L/L$ vs. H, for $Ni_2MnGa_{0.88}Cu_{0.12}$ up to 9.8 T. The arrows show the process of the magnetic field.

In [42] and [43], regarding $Ni_2MnGa_{1-x}Cu_x$, the magnetic field was conceived to shrink because of the x dependence of Cu concentration of the unit-cell volume within the interval $0 \leq x \leq 0.25$ at room temperature. The unit-cell volume decreases with increasing x. $Ni_2MnGa_{1-x}Cu_x$ in the concentration range of $0 \leq x \leq 0.07$ crystallize in the $L2_1$ cubic structure at room temperature. The lattice parameter a_{L21} with $x = 0.02$ ($L2_1$ cubic structure) was 0.58206 nm. For $0.10 \leq x \leq 0.25$, the sample crystallizes as a $D0_{22}$-like martensite structure, as mentioned in Section 2.1. At $x = 0.12$, the unit-cell volume defined through X-ray diffraction was smaller than that of the extrapolation line at $0 \leq x \leq 0.07$ ($L2_1$ cubic structure) of the order of 0.200% (2000 ppm). The linear strain of a polycrystal is one-third of the volume strain [44]. Therefore, 670-ppm shrinkage is expected because of the austenite phase transition to the martensite phase. Moreover, in the martensite phase, the magnetic easy axis is a_{D022}, and a large shrinkage was observed with the applied magnetic field parallel to the easy axis [15]. Therefore, negative magnetostriction could be observed around T_M for this study.

3.3. Comparation with Other Ferromagnetic Magneto-Structural Alloys

Now, we compare the magneto-structural properties of $Ni_2MnGa_{0.88}Cu_{0.12}$ with other Ni_2MnGa-type ferromagnetic alloys, i.e., Ni_2MnGa [17], $Ni_{41}Co_9Mn_{31.5}Ga_{18.5}$ [31], $Ni_2Mn_{0.7}Cu_{0.3}Ga_{0.84}Al_{0.16}$ [19], and $Ni_{2.15}Mn_{0.70}Cr_{0.15}Ga$ [20] polycrystalline samples, as well as $Ni_{45}Co_5Mn_{36.7}In_{13.3}$ [7] and $Ni_{50}Mn_{30}Ga_{20}$ [5] single crystalline samples.

Table 1 provides information on the crystal structure in the martensite phase, volume change from the martensite phase to the austenite phase, and magnetostriction value of varioys Ni_2MnGa-type Heusler alloys.

Table 1. Crystal structure, volume change, and linear magnetostriction $(\Delta L/L)_{//}$ value of the Ni_2MnGa-type Heusler alloys. All alloys have the $L2_1$ cubic crystal structure in the parent austenite phase. Arrows indicate the transition direction of the magnetic and crystallographic phases.

Alloy	Crystal Structure (Martensite Phase)	Volume Change (ppm) (Paramagnetic Austenite →Ferromagnetic Martensite)	Linear Magnetostriction ($\Delta L/L$(Paramagnetic Austenite →Ferromagnetic Martensite)
Ni_2MnGa	*14M* [1]	−330 [1]	−780 [2]
$Ni_2MnGa_{0.88}Cu_{0.12}$ [3]	$D0_{22}$ + *14M*	−2000	−1300 (this study)
$Ni_{41}Co_9Mn_{31.5}Ga_{18.5}$ [4]	tetragonal	5700	1900
$Ni_{45}Co_5Mn_{36.7}In_{13.3}$ [5]	*14M*	---	30,000 (single crystal)
$Ni_2Mn_{0.7}Cu_{0.3}Ga_{0.84}Al_{0.16}$ [6]	$L1_0$	−55,000	−26,000
$Ni_{2.15}Mn_{0.70}Cr_{0.15}Ga$ [7]	$L1_0$	−30,000	−8100
$Ni_{50}Mn_{30}Ga_{20}$ [8]	*5M*	−260	−3300 (single crystal) (this study)

[1] Ref. [45] [2] Ref. [17] [3] Ref. [37] [4] Ref. [31] [5] Ref. [7] [6] Ref. [19] [7] Ref. [20] [8] Ref. [5]. The arrows indicate the transition direction of the magnetic and crystallographic phases.

Sakon et al. investigated the magnetostriction of polycrystalline sample of Ni_2MnGa [17]. The magnetostriction was measured at T = 185 K in the martensite phase, where the temperature was just below T_M = 193 K. The magnetostriction was measured parallel $(\Delta L/L)_{//}$ and perpendicular $(\Delta L/L)_{\perp}$ to the magnetic field. Furthermore, the obtained magnetostriction values were $(\Delta L/L)_{//} = -780$ ppm and $(\Delta L/L)_{\perp}$ = 260 ppm. The volume magnetostriction $\Delta V/V$ can be obtained as follows [44–46]:

$$\Delta V/V = (\Delta L/L)_{//} + 2\,(\Delta L/L)_{\perp} \quad (1)$$

The obtained $\Delta V/V$ was −260 ppm. Singh et al. [45] obtained the temperature dependencies of the lattice constant and unit-cell volume based on a high-resolution synchrotron X-ray powder diffraction. The volume change from a $L2_1$ cubic structure to a monoclinic *14M* structure was −330 ppm. Volume magnetostriction $\Delta V/V$ was 80% of the volume change of the $L2_1 \rightarrow$ *14M* martensite transition.

$Ni_{41}Co_9Mn_{31.5}Ga_{18.5}$ has a T_M of 320 K and a Curie temperature, T_C of 468 K [31]. The $L2_1$ cubic ferromagnetic austenite structure and tetragonal ferrimagnetic martensite structure were obtained as magnetic and crystalline structure, respectively. Thermo-magnetic *M–T* curves indicated that at 0.5 T, a sharp magnetization jump corresponding to the thermal martensitic transition temperatures [martensite → austenite T_R = 350 K (heating process) and austenite → martensite T_M = 320 K (cooling process)] was observed at around room temperature. The T_M decreased with increasing magnetic field at the ratio of $dT_M/d(\mu_0H) = -12.6$ K/T. The absolute value of $dT_M/d(\mu_0H)$ of $Ni_{41}Co_9Mn_{31.5}Ga_{18.5}$ is 10 times larger than that of $Ni_2MnGa_{0.88}Cu_{0.12}$. Below T_M, metamagnetic transitions were observed for the magnetization processes (*M-T* curves). In addition, magnetostrictions larger than 1000 ppm were observed at the temperature of 150 K $\leq T \leq$ 290 K below T_M, corresponding to a metamagnetic transition. At 290 K, the total value of magnetostriction was 1900 ppm. Assuming isotropic magnetostriction for this polycrystalline sample, the total volume change can be estimated as 5700 ppm [31]. These results indicate that below T_M, a reverse martensite transition occurred from the ferrimagnetic martensite phase to the ferromagnetic austenite phase with the increasing magnetic field. The experimental and theoretical results show that the total energy of the ferromagnetic state is close to that of the ferrimagnetic martensite state, and the magnetic-field-induced ferromagnetic austenite phase is unstable in the ratio of the tetragonality (*c/a*) for 1.0 $\leq c/a \leq$ 1.1. Therefore, metamagnetic transitions and large magnetostrictions were observed in the magnetic field.

A single crystal of $Ni_{45}Co_5Mn_{36.7}In_{13.3}$ indicates a large MFIS of 30,000 ppm (3.0%) at 298 K when a compressive pre-strain of approximately 3% was applied in the direction plotted with a filled circle in the stereographic triangle, as shown in Figure 4 of Ref. [7]. The magnetic field was applied parallel to the compressive axis of the sample. The T_M was 290 K and T_C was 382 K. Below this T_M, the *M-T* curve indicated a clear metamagnetic transition behavior. The T_M decreased with increasing magnetic field at the ratio of $dT_M/d(m_0H) = -7$ K/T, and an $L2_1$ cubic structure was achieved in austenite phase, and a monoclinic *14 M* layered structure was achieved in the martensite phase. Below T_M, the MFIS was observed to occur during the metamagnetic transition from the ferrimagnetic martensite phase to ferromagnetic austenite phase. The strain from the $L2_1$ cubic structure to monoclinic *14 M* layered structure is significant, exceeding a few percent. For example, Sozinov et al. experimentally studied the MFIS of a single-variant sample of an orthorhombic seven-layered phase in the $Ni_{48.8}Mn_{29.7}Ga_{21.5}$ single crystalline alloy at 300 K, measuring it perpendicular to the magnetic field applied along the [100] direction [3]. They observed a giant MFIS of approximately 95,000 ppm (9.5%) at an ambient temperature in a magnetic field of less than 1 T in the Ni_2MnGa orthorhombic seven-layered (*7M* structure, which corresponds to *14M* structure) martensite phase.

On the contrary, the crystal structure of $Ni_{41}Co_9Mn_{31.5}Ga_{18.5}$ is not a monoclinic *14 M* structure but a tetragonal structure, and the volume change due to the martensitic transition is 5700 ppm. Therefore, a moderate magnitude magnetostriction was observed.

In the case of $Ni_2MnGa_{0.88}Cu_{0.12}$, the value of the magnetostriction (1300 ppm) is the same order as that of $Ni_{41}Co_9Mn_{31.5}Ga_{18.5}$ (1900 ppm). The tetragonal phase of the $D0_{22}$-like crystal structure coexists with the monoclinic phase of the monoclinic *14M* structure (space group: *P2/m*) in the martensite phase at room temperature. The volume change from the $L2_1$ cubic structure to monoclinic *14M* structure was −330 ppm for Ni_2MnGa [45]. Therefore, −1300 ppm magnetostriction, which is smaller than that of $Ni_{41}Co_9Mn_{31.5}Ga_{18.5}$, the crystal structure of which is a tetragonal martensite, is suitable for the $Ni_2MnGa_{0.88}Cu_{0.12}$ polycrystalline alloy.

As mentioned earlier, for $Ni_2Mn_{0.7}Cu_{0.3}Ga_{0.84}Al_{0.16}$, $T_M = 293$ K and the martensitic and ferromagnetic transitions occur at the same temperature, as confirmed by the magnetization and magnetostriction measurements [19]. In addition, the results of X-ray powder diffraction measurement indicated that the crystal structure was $L2_1$ cubic austenite at 312 K. Moreover, at 298 K, which is slightly higher than $T_M = 293$ K, the austenitic $L2_1$ cubic structure and martensitic $L1_0$ non-modulated tetragonal structure were observed to coexist. Finally, below 257 K, the $L1_0$ structure was achieved.

The unit-cell volume of $Ni_2Mn_{0.7}Cu_{0.3}Ga_{0.84}Al_{0.16}$ calculated by the lattice parameters showed the volume difference between the austenite $L2_1$ phases and martensite $L1_0$ phase as approximately −55,000 ppm (−5.5%). Therefore, a giant magnetostriction was expected around −18,000 ppm (−1.8%) for the polycrystalline sample. However, the experimental results showed magnetostriction of $\Delta L/L = -26{,}000$ ppm (−2.6%.), which is larger than the expected value. Therefore, the effect of the alignment of the variant of martensite must be considered. The magnetostriction is irreversible during the experimental temperatures of $297.4\ \mathrm{K} \leq T \leq 304$ K. Partial recovery was observed only at 305 K.

In the case of $Ni_2MnGa_{0.88}Cu_{0.12}$, the value of the magnetostriction was one-tenth that of $Ni_2Mn_{0.7}Cu_{0.3}Ga_{0.84}Al_{0.16}$. However, an 85% recovery strain was observed at 345 K. The existence of recovery strain at atmospheric pressure is advantageous for the alloy's application in sensors and actuators.

Mendonça et al. [20] performed the magnetostriction measurements on the $Ni_{2.15}Mn_{0.70}Cr_{0.15}Ga$ polycrystalline alloy with $T_M = 299$ K. In addition, the martensitic and ferromagnetic transitions occurred at the same temperature, as confirmed by the magnetization and magnetostriction measurements. The resulting crystal structure in the parent austenite phase was the $L2_1$ cubic structure. The martensite phase is an $L1_0$ non-modulated tetragonal structure. The unit-cell volume of $Ni_2Mn_{0.7}Cu_{0.3}Ga_{0.84}Al_{0.16}$ calculated from the lattice parameters showed that the volume difference between the austenite $L2_1$ phases and martensite $L1_0$ phase is approximately −30,000 ppm (−3.0%). The magnetostriction results indicated a reversible strain around −8100 ppm (−0.81%) under 0–9 T, which is slightly lesser than one-third the volume difference (−10,000 ppm, −1.0%). Mendonça et al. [20] overcame the weak point of large hysteresis in the magnetostriction of $Ni_2Mn_{0.7}Cu_{0.3}Ga_{0.84}Al_{0.16}$. The characteristic of a magneto-structural transformation at room temperature with low hysteresis is advantageous for the alloy's application in functional magnetic materials. In this experimental study, the value of the magnetostriction of $Ni_2MnGa_{0.88}Cu_{0.12}$ is one-sixth of that of $Ni_{2.15}Mn_{0.70}Cr_{0.15}Ga$. However, it is advantageous for its industrial application and possesses smaller magnetostriction than that of $Ni_{2.15}Mn_{0.70}Cr_{0.15}Ga$.

As a reference sample, we measured the magnetostriction of the Ni_2MnGa-type ($Ni_{50}Mn_{30}Ga_{20}$) alloy manufactured by Adaptamat Co., Ltd. [4,5], which is shown in Figure 6. This alloy causes martensite phase transition at T_M = 315 K. Magnetostriction of $Ni_{50}Mn_{30}Ga_{20}$ was measured parallel to the magnetic field, which was parallel to the c-axis. Magnetostriction of −3300 ppm was observed at an atmospheric pressure and 298 K. Almost 100% recovery strain was observed. The value of the magnetostriction of $Ni_2MnGa_{0.88}Cu_{0.12}$ was one-third that of $Ni_{50}Mn_{30}Ga_{20}$. However, the value of the magnetostriction of $Ni_2MnGa_{0.88}Cu_{0.12}$ is larger than that of Terfenol-D (800 ppm), which is renowned as the most commonly used magnetostrictive alloy [47,48].

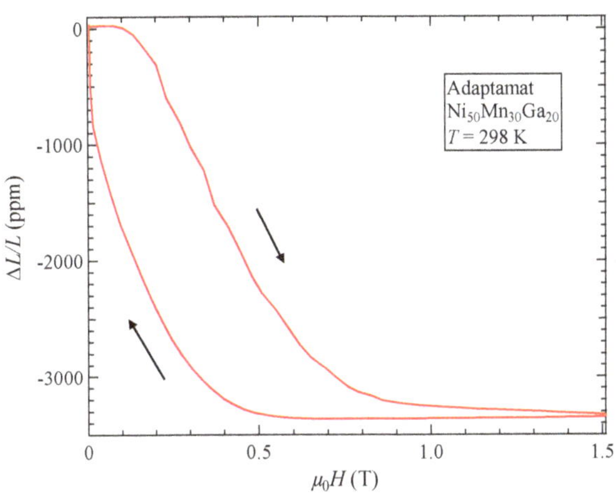

Figure 6. Magnetic field dependence of forced linear magnetostriction, Δ*L*/*L* vs. *H*, for $Ni_{50}Mn_{30}Ga_{20}$. The arrows show the process of the magnetic field.

As mentioned above, a shrinkage of −3300 ppm was observed at 298 K in the martensite phase. This result is comparable to magnetostriction of Ni_2MnGa single crystal at atmospheric pressure (unstressed crystal) measured by Ullakko et al. [1]. The martensitic transition temperature, T_M, was 275 K. In the martensite phase at 265 K, a magnetostriction value of the [001] axis (*c*-axis in a martensite phase) strain in response to magnetic field applied along *c*-axis was—1000 ppm in an applied magnetic field of 1 T. Ullakko et al. mentioned that the strain occurs fully within the martensitic phase and is due to the motion of twin boundaries. The magnetostriction is only a small fraction of the lattice constant change (Δ*c*/*c* = 6.56% = 65,600 ppm). They also mentioned that this small strain is caused by the strain accommodation through different twin variant orientations. In the martensite phase, applying the magnetic field to the single variant material causes the other twin variants to appear and grow [3,4]. When magnetic field strength increases the boundaries between twins move, and the preferentially oriented twin variants grow at the expense of the other twin variants. It is interesting that the magnetostriction of $Ni_{50}Mn_{30}Ga_{20}$ (this study) and Ni_2MnGa [1] was almost fully reversible. Currently, we cannot explain why the recovery of strain occurred in the martensite phase. It is conceived that the rearrangements of the twin variants would occur when increasing a magnetic field, and afterwards, during the decreasing magnetic field process, elastic recovery may act and come back to the original state. However, the exact reason for this recovery is currently not clear. Recently, Wu et al. simulated the ferromagnetic domain structure and martensite variant microstructure of Ni_2MnGa-type shape-memory alloy [12]. They found −0.28% (−2800 ppm) magnetic field induced strain and also found 100% recovery strain at 285 K below T_M = 300 K. This value is comparable with our result of $Ni_{50}Mn_{30}Ga_{20}$ alloy. We retrieve this model helps explanation of the recovery strain.

4. Conclusions

In this study, magnetostriction measurements were performed on the ferromagnetic Heusler alloy $Ni_2MnGa_{0.88}Cu_{0.12}$. One characteristic of this alloy is that its martensitic and ferromagnetic transitions are caused at the same temperature. In the austenite and martensite phases, the alloy crystallizes with $L2_1$ and $D0_{22}$-like crystal structures, respectively. Metamagnetic transition was observed in the magnetic field of $\mu_0 H$ = 4 T at 344 K. This result showed that the occurrence of phase transition was induced by the magnetic field under constant temperature. Forced magnetostriction measurements ($\Delta L/L$) were performed under constant temperature and at an atmospheric pressure of P = 0.1 MPa. Magnetostriction up to 1300 ppm was observed around the T_M. The magnetization and magnetostriction measurement results showed that the magnetic-field-induced strain from the paramagnetic austenite phase to ferromagnetic martensite phase occurred. In the case of $Ni_2MnGa_{0.88}Cu_{0.12}$, the value of the magnetostriction was one-tenth that of $Ni_2Mn_{0.7}Cu_{0.3}Ga_{0.84}Al_{0.16}$, which has a tetragonal $L1_0$ martensite structure. However, 85% recovery strain was observed at 345 K. The value of the magnetostriction of $Ni_2MnGa_{0.88}Cu_{0.12}$ was one-sixth that of $Ni_{2.15}Mn_{0.70}Cr_{0.15}Ga$, which has low hysteresis and reversible strain in magnetic fields. The existence of a recovery strain at atmospheric pressure is advantageous to the alloy's application to sensors and actuators.

For reference, we measured the magnetostriction of the Ni_2MnGa-type ($Ni_{50}Mn_{30}Ga_{20}$) alloy made in Adaptamat Co., Ltd. The magnetostriction of −3300 ppm was observed at 298 K, and almost 100% recovery strain was observed. The value of the magnetostriction of $Ni_2MnGa_{0.88}Cu_{0.12}$ was one-third that of $Ni_{50}Mn_{30}Ga_{20}$. However, the value of the magnetostriction of $Ni_2MnGa_{0.88}Cu_{0.12}$ is larger than that of Terfenol-D (800 ppm).

Author Contributions: Y.A. and T.K. prepared the samples; T.S., H.N., Y.N., M.H. and T.K. conceived and designed the experiments; K.M., T.K., H.N., Y.N., M.H. and T.S. performed the experiments; K.M. and T.S. analyzed the data; T.S., K.M., T.K., Y.N., M.H., H.N. and Y.A. wrote the paper. All authors have read and agreed to the published version of the manuscript.

Funding: This research received no external funding.

Data Availability Statement: Not applicable.

Acknowledgments: The authors express their gratitude to K. Endo for their contribution to sample preparation. This research was carried out at the Center for Advanced High Magnetic Field Science in Osaka University under the Visiting Researchers' Program of the Institute for Solid State Physics, the University of Tokyo and at the High Field Laboratory for Superconducting Materials, Institute for Materials Research, Tohoku University.

Conflicts of Interest: The authors declare no conflict of interest.

References

1. Ullakko, K.; Huang, J.K.; Kantner, C.; O'Handley, R.C.; Kokorin, V.V. Large magnetic-field-induced strains in Ni_2MnGa single crystals. *Appl. Phys. Lett.* **1996**, *69*, 1966–1968. [CrossRef]
2. Ullakko, K.; Huang, J.K.; Kokorin, V.V.; O'Handley, R.C. Magnetically controlled shape memory effect in Ni_2MnGa intermetallics. *Scr. Mater.* **1997**, *36*, 1133–1138. [CrossRef]
3. Sozinov, A.; Likhachev, A.A.; Lanska, N.; Ullakko, K. Giant magnetic-field-induced strain in NiMnGa seven-layered martensitic phase. *Appl. Phys. Lett.* **2002**, *80*, 1746–1748. [CrossRef]
4. Tellinen, J.; Suorsa, I.; Jääskeläinen, A.; Aaltio, I.; Ullakko, K. Basic properties of magnetic shape memory actuators. In Proceedings of the 8th International Conference ACTUATOR, Bremen, Germany, 10–12 June 2002; pp. 566–569.

5. Ge, Y.; Söderberg, O.; Lanska, N.; Sozinov, A.; Ullakko, K.; Lindroos, V.K. Crystal structure of three Ni-Mn-Ga alloys in powder and bulk materials. *J. Phys. IV* **2003**, *112*, 921–924. [CrossRef]
6. Chernenko, V.A.; L'vov, V.A. Magnetoelastic nature of ferromagnetic shape memory effect. *Mater. Sci. Forum* **2008**, *583*, 1–20. [CrossRef]
7. Kainuma, R.; Imano, Y.; Ito, W.; Sutou, Y.; Morito, H.; Okamoto, S.; Kitakami, O.; Oikawa, K.; Fujita, A.; Kanomata, T.; et al. Magnetic-field-induced shape recovery by reverse phase transformation. *Nature* **2006**, *439*, 957–960. [CrossRef]
8. Sakon, T.; Sasaki, K.; Numakura, D.; Abe, M.; Nojiri, H.; Adachi, Y.; Kanomata, T. Magnetic field-induced transition in co-doped $Ni_{41}Co_9Mn_{31.5}Ga_{18.5}$ heusler alloy. *Mater. Trans.* **2013**, *54*, 9–13. [CrossRef]
9. Karaca, H.E.; Karaman, I.; Basaran, B.; Ren, Y.; Chumlyakov, Y.I.; Maier, H.J. Magnetic field-induced phase transformation in NiMnCoIn magnetic shape-memory alloys—A new actuation mechanism with large work output. *Adv. Funct. Mater.* **2009**, *19*, 983–998. [CrossRef]
10. Kainuma, R.; Oikawa, K.; Ito, W.; Sutou, Y.; Kanomata, T.; Ishida, K. Metamagnetic shape memory effect in NiMn-based Heusler-type alloys. *J. Mater. Chem.* **2008**, *18*, 1837. [CrossRef]
11. O'Handley, R.C.; Murray, S.J.; Marioni, M.; Nembach, H.; Allen, S.M. Phenomenology of giant magnetic-field-induced strain in ferromagnetic shape memory materials. *J. Appl. Phys.* **2000**, *87*, 4712–4717. [CrossRef]
12. Wu, P.; Liang, Y. Enhanced reversible magnetic-field-induced strain in Ni-Mn-Ga Alloy. *Metals* **2021**, *11*, 2017. [CrossRef]
13. Kumar, A.S.; Seshubai, V. 0.7% magnetic field induced strain in polycrystalline $Ni_{50}Mn_{29}Ga_{21}$ ferromagnetic shape memory alloy. *Int. J. Innov. Res. Sci. Eng. Technol.* **2013**, *2*, 4226–4232.
14. Mennerich, C.; Wendler, F.; Jainta, M.; Nestler, B. Rearrangement of martensitic variants in Ni_2MnGa studied with the phase-field method. *Eur. Phys. J. B* **2013**, *86*, 171. [CrossRef]
15. Okamoto, N.; Fukuda, T.; Kakeshita, T.; Takeuchi, T.; Kishino, K. Rearrangement of variants in Ni_2MnGa under magnetic field. *Sci. Technol. Adv. Mater.* **2004**, *5*, 29–34. [CrossRef]
16. Sakon, T.; Fujimoto, N.; Kanomata, T.; Adachi, Y. Magnetostriction of $Ni_2Mn_{1-x}Cr_xGa$ heusler alloys. *Metals* **2017**, *7*, 410. [CrossRef]
17. Sakon, T.; Yamasaki, Y.; Kodama, H.; Kanomata, T.; Nojiri, H.; Adachi, Y. The characteristic properties of magnetostriction and magneto-volume effects of Ni_2MnGa-type ferromagnetic heusler alloys. *Materials* **2019**, *12*, 3655. [CrossRef] [PubMed]
18. Sakon, T.; Adach, Y.; Kanomata, T. Magneto-structural properties of Ni_2MnGa ferromagnetic shape memory alloy in magnetic fields. *Metals* **2013**, *3*, 202–224. [CrossRef]
19. Mendonça, A.A.; Jurado, J.F.; Stuard, S.J.; Silva, L.E.L.; Eslava, G.G.; Cohen, L.F.; Ghivelder, L.; Gomes, A.M. Giant magnetic-field-induced strain in Ni_2MnGa-based polycrystal. *J. Alloys Compd.* **2018**, *738*, 509–514. [CrossRef]
20. Mendonça, A.A.; Ghivelder, L.; Bernardo, P.L.; Cohen, L.F.; Gomes, A.M. Low hysteretic magnetostructural transformation in Cr-doped Ni-Mn-Ga Heusler alloy. *J. Alloys Compd.* **2023**, *938*, 168444. [CrossRef]
21. Sofronie, M.; Tolea, F.; Tolea, M.; Popescu, B.; Valeanu, M. Magnetic and magnetostrictive properties of the ternary $Fe_{67.5}Pd_{30.5}Ga_2$ ferromagnetic shape memory ribbons. *J. Phys. Chem. Sol.* **2020**, *142*, 109446. [CrossRef]
22. Mahfouzi, M.; Carman, G.P.; Kioussis, N. Magnetoelastic and magnetostrictive properties of Co2XAl Heusler compounds. *Phys. Rev. B* **2020**, *102*, 094401. [CrossRef]
23. Sofronie, M.; Tolea, M.; Popescu, B.; Enculescu, M.; Tolea, F. Magnetic and Magnetostrictive Properties of $Ni_{50}Mn_{20}Ga_{27}Cu_3$ Rapidly Quenched Ribbons. *Materials* **2021**, *14*, 5126. [CrossRef]
24. Liu, K.; Ma, S.; Zhang, Y.; Zeng, H.; Yu, G.; Luo, X.; Chena, C.; Rehman, S.U.; Hu, Y.; Zhong, Z. Magnetic-field-driven reverse martensitic transformation with multiple magneto-responsive effects by manipulating magnetic ordering in Fe-doped Co-V-Ga Heusler alloys. *J. Mater. Sci. Technol.* **2020**, *58*, 145–154. [CrossRef]
25. Kaštil, J.; Kamarád, J.; Míšek, J.; Isnard, O.; Amara, M.; Arnold, Z. Magnetostriction and extraordinary exchange spring and exchange bias effects in $Ni_{48}Mn_{39}Sn_{13}$ Heusler alloy. *Intermetallics* **2021**, *132*, 109137. [CrossRef]
26. Huang, Y.; Qian, J.; Dong, D.; Shi, Y.; Du, Y.; Tang, S. Magnetostriction in <0kl>-oriented composites with CoMnSi microspheres. *J. Magn. Magn. Mater.* **2022**, *543*, 168621.
27. Liua, J.; Gonga, Y.; Zhang, F.; Youa, Y.; Xu, G.; Miao, X.; Xu, F. Large, low-field and reversible magnetostrictive effect in MnCoSi-based metamagnet at room temperature. *J. Mater. Sci. Technol.* **2021**, *76*, 104–110. [CrossRef]
28. Minorowicz, B.; Milecki, A. Design and Control of Magnetic Shape Memory Alloy Actuators. *Materials* **2022**, *15*, 4400. [CrossRef]
29. Kurita, H.; Keino, T.; Senzaki, T.; Narita, F. Direct and inverse magnetostrictive properties of Fe–Co–V alloy particle-dispersed polyurethane matrix soft composite sheets. *Sens. Actuators A Phys.* **2022**, *337*, 113427. [CrossRef]
30. Sadeghzadeh, A.; Asua, E.; Feuchtwanger, J.; Etxebarria, V.; García-Arribas, A. Ferromagnetic shape memory alloy actuator enabled for nanometric position control using hysteresis compensation. *Sens. Actuators A Phys.* **2012**, *182*, 122–129. [CrossRef]
31. Kihara, T.; Roy, T.; Xu, X.; Miyake, A.; Tsujikawa, M.; Mitamura, H.; Tokunaga, M.; Adachi, Y.; Eto, T.; Kanomata, T. Observation of inverse magnetocaloric effect in magnetic-field-induced austenite phase of Heusler alloys Ni50−xCoxMn31.5Ga18.5 (x = 9 and 9.7). *Phys. Rev. Mater.* **2021**, *5*, 034416. [CrossRef]
32. Hennel, M.; Galdun, L.; Džubinská, A.; Reiffers, M.; Varga, R. High efficiency direct magnetocaloric effect in Heusler Ni_2MnGa microwire at low magnetic fields. *J. Alloys Compd.* **2023**, *960*, 170621. [CrossRef]
33. Brock, J.; Khan, M. Large refrigeration capacities near room temperature in $Ni_2Mn_{1-x}Cr_xIn$. *J. Magn. Magn. Mater.* **2017**, *425*, 1–5. [CrossRef]

34. Zheng, T.; Liu, K.; Chen, H.; Wang, C. Large magnetocaloric and magnetoresistance effects during martensitic transformation in Heusler-type $Ni_{44}Co_6Mn_{37}In_{13}$ alloy. *J. Magn. Magn. Mater.* **2022**, *563*, 170034. [CrossRef]
35. Salazar-Mejía, C.; Devi, P.; Singh, S.; Felser, C.; Wosnitza, J. Influence of Cr substitution on the reversibility of the magnetocaloric effect in Ni-Cr-Mn-In Heusler alloys. *Phys. Rev. Mater.* **2021**, *5*, 104406. [CrossRef]
36. Kitanovski, A. Energy applications of magnetocaloric materials. *Adv. Energy Mater.* **2020**, *10*, 1903741. [CrossRef]
37. Sakon, T.; Nagashio, H.; Sasaki, K.; Susuga, S.; Numakura, D.; Abe, M.; Endo, K.; Nojiri, H.; Kanomata, T. Thermal expansion and magnetization studies of the novel ferromagnetic shape memory alloy $N_2MnGa_{0.88}Cu_{0.12}$ in a magnetic field. *Phys. Scr.* **2011**, *84*, 045603. [CrossRef]
38. Khovailo, V.V.; Takagi, T.; Tani, J.; Levitin, R.Z.; Cherechukin, A.A.; Matsumoto, M.; Note, R. Magnetic properties of Ni2.18Mn0.82Ga heusler alloys with a coupled magnetostructural transition. *Phys. Rev. B* **2002**, *65*, 092410. [CrossRef]
39. Filippov, D.A.; Khovailo, V.V.; Koledov, V.V.; Krasnoperov, E.P.; Levitin, R.Z.; Shavrov, V.G.; Takagi, T. The magnetic field influence on magnetostructural phase transition in $Ni_{2.19}Mn_{0.81}Ga$. *J. Magn. Magn. Mater.* **2003**, *258*, 507–509. [CrossRef]
40. Khovailo, V.V.; Novosad, V.; Takagi, T.; Filippov, D.A.; Levitin, R.Z.; Vasil'ev, A.N. Magnetic properties and magnetostructural phase transitions in shape memory alloys. *Phys. Rev. B* **2004**, *70*, 174413. [CrossRef]
41. Gonzàlez-Comas, A.; Obradó, E.; Mañosa, L.; Planes, A.; Chernenko, V.A.; Hattink, B.J.; Labarta, A. Premartensitic and martensitic phase transitions in ferromagnetic Ni_2MnGa. *Phys. Rev. B* **1999**, *60*, 7085–7090. [CrossRef]
42. Endo, K.; Kanomata, T.; Kimura, A.; Kataoka, M.; Nishihara, H.; Umetsu, R.Y.; Obara, K.; Shishido, T.; Nagasako, M.; Kainuma, R.; et al. Magnetic phase diagram of the ferromagnetic shape memory alloys $Ni_2MnGa_{1-x}Cu_x$. *Mater. Sci. Forum* **2011**, *684*, 165–176. [CrossRef]
43. Kanomata, T.; Endo, K.; Kudo, N.; Umetsu, R.Y.; Nishihara, H.; Kataoka, M.; Nagasako, M.; Kainuma, R.; Ziebeck, K.R.A. Magnetic moment of Cu-modified Ni_2MnGa magnetic shape memory alloys. *Metals* **2013**, *3*, 114–122. [CrossRef]
44. Kittel, C. *Introduction of Solid State Physics*, 8th ed.; John Wiley & Sons Inc.: Hoboken, NJ, USA, 2004; p. 75.
45. Singh, S.; Bednarcik, J.; Barman, S.R.; Felser, C.; Pandey, D. Premartensite to martensite transition and its implications for the origin of modulation in Ni_2MnGa ferromagnetic shape memory alloy. *Phys. Rev. B* **2015**, *92*, 054112. [CrossRef]
46. Nizhankovskii, V.I. Classical magnetostriction of nickel in high magnetic field. *Eur. Phys. J. B* **2006**, *53*, 1–4. [CrossRef]
47. Wang, Z.; Liu, J.; Jiang, C.; Xu, H. The stress dependence of magnetostriction hysteresis in TbDyFe [110] oriented crystal. *J. Appl. Phys.* **2011**, *109*, 123923. [CrossRef]
48. Sakon, T.; Matsumoto, T.; Komori, T. Rotation angle sensing system using magnetostrictive alloy Terfenol-D and permanent magnet. *Sens. Actuators A* **2021**, *321*, 112588. [CrossRef]

metals

MDPI

Article

Terahertz Metamaterial Absorber Based on Ni–Mn–Sn Ferromagnetic Shape Memory Alloy Films

Rui Liu [1], Xiaochuan Wang [2], Jiachen Zhu [3], Xiaohua Tian [4], Wenbin Zhao [2], Changlong Tan [2,*] and Kun Zhang [5,*]

1 School of Science, Harbin University of Science and Technology, Harbin 150080, China
2 School of Materials Science and Chemical Engineering, Harbin University of Science and Technology, Harbin 150080, China
3 School of Materials Science and Engineering, Harbin Institute of Technology, Harbin 150001, China
4 School of Electrical and Electronic Engineering, Harbin University of Science and Technology, Harbin 150080, China
5 Shenyang National Laboratory for Materials Science, Institute of Metal Research, Chinese Academy of Sciences, Shenyang 110016, China
* Correspondence: changlongtan@hrbust.edu.cn (C.T.); kzhang@imr.ac.cn (K.Z.)

Abstract: Terahertz absorbers have been extensively investigated by researchers due to their applications in thermophotovoltaic energy conversion and sensors, but a key factor limiting their development is the lack of vital and versatile materials. Ferromagnetic shape memory alloys (FSMAs) offer a novel remedy for tunable metamaterials due to their brilliant recovery of deformation, remote control, and transient response. In this study, we propose a tunable absorber based on magnetic field tuning, consisting of Ni–Mn–Sn ferro-magnetic shape memory alloy films in fractal geometry and optically excited Si plates. Numerical analysis shows that the proposed absorber has an absorbance bandwidth of 1.129 THz above 90% between 1.950 THz and 3.079 THz. The absorber geometry can be regulated by an external magnetic field, allowing dynamic switching between broadband and narrowband absorption modes, the latter showing an ultra-narrow bandwidth and a high-quality factor Q of ~25.8. The proposed terahertz absorber has several advantages over current state-of-the-art bifunctional absorbers, including its ultra-thin structure of 10.39 μm and an additional switching function. The absorption can be continuously tuned from 90% to 5% when the light-excited silicon plate is transferred from the insulator state to the metal state. This study presents a promising alternative strategy for developing actively regulated and versatile terahertz-integrated devices.

Keywords: metamaterials; shape memory alloy films; Ni–Mn–Sn alloy films; terahertz absorber; dynamic tuning

1. Introduction

Based on the distinctive advantages of their photo energy and bandwidth, terahertz (THz) frequency waves have great potential for broad application in safety inspection, explosive detection, spectroscopy, medical imaging, wireless communications, etc. [1–5]. Tremendous research has been conducted in designing high-performance THz devices for the development of miniaturized components. Metamaterials (MMs) have recently emerged as perfect candidates for manipulating THz waves, attributable to their unique electromagnetic response mechanism [6–8]. Among them, THz metamaterial absorbers (TMMAs) are highly desired in a broad range of applications, including high-efficiency photodetectors, THz imaging, and stealth technology [9–11].

Research on TMMAs can be traced back to the concept of a perfect metamaterial absorber in the microwave range proposed by Landy et al. [12]. Since then, it has opened the door to developing TMMAs worldwide. Initially, conventional TMMAs consisted of a typical sandwich structure with a dielectric spacer, and these TMMAs showed narrowband absorption [13–15]. To enrich the functionalities, TMMAS with dual-, multi-, and broadband absorption have been proposed and investigated in recent years [16–18]. Achieving broadband absorption of terahertz waves is still a challenge compared to visible and infrared light. In order to achieve broadband absorption, researchers have made many attempts, such as increasing the number of multiple distinct circular metallic patches, composite multi-layer stacking (vertical arrangement), introducing lumped elements, and the metal-dielectric nanocomposite systems [19–23], but these methods, although capable of achieving broadband absorption properties, also lead to a high cost of the proposed absorber, complex fabrication process, large size, and other drawbacks, which limit their practical applications. The above problem can be solved by using a fractal structure, which is a self-similar repetitive geometric structure in a fixed arrangement with a self-similar pattern leading to multiple resonant coupling, thus achieving broadband absorption [24–27]. More recently, the Cayley tree fractal meta-resonator supercells were proposed by Qasim Mehmood et al. [28]. They integrated different Cayley tree fractal resonators into one supercell based on the frequency shifting and multi-resonance bands of different fractal orders, which realizes broadband, polarization-insensitive ultrathin monolayer terahertz metamaterial absorbers. While broadband absorbers are mostly used in thermo-photovoltaics, narrowband perfect absorbers can be used in sensing, absorption filtering, and thermal radiation tailoring [29–32]. Most of the designs in the literature serve as either broadband or narrowband absorbers, and only a few studies achieve double functionality [33–41]. Combining both functionalities in the same device remains a significant challenge. Moreover, in conventional TMMAs, the resonance performance is determined by the structure parameters and material properties. Once fabricated, the THz absorber has non-adjustable absorption performance limiting its application. Although there have been studies on tunable materials with improved performance of tunable broadband terahertz metamaterial absorbers, the dilemma of multifunctional performance versus the complex multilayer structure of tunable TMMAs remains a problem. Therefore, designing thin and simple THz absorbers with tunable absorption performance remains highly desirable.

Different from the traditional metallic arrays of TMMAs, shape memory alloys (SMAs) exhibit distinctive functional characteristics, including superelastic behavior, shape memory effect, and work output with a high power/weight ratio [42,43]. Designing artificial metamaterial composed of shape memory alloys provides an active modulation by utilizing recoverable deformation and phase transformation in the subwavelength structures. This active adjustment has a bigger regulation region and reduces the complexity of the structure. In our previous works, Ni–Ti SMAs were innovatively introduced into the development of THz metamaterial devices [44,45]. Nevertheless, the slow martensitic transition response of Ni–Ti SMA is due to temperature-driven, magnetic-field-induced SMAs, which have attracted extensive research as a magnetoactuator material [46–48]. The controlling mode by a magnetic field is more flexible and quicker. Based on the FSMAs, it can be expected to achieve the dynamic tunability of terahertz metamaterial absorbers, which show the advantages of fast response and non-contact control.

In this paper, we propose a multifunctional, ultrathin, and high-performance TMMA by combining the horizontally aligned Caley tree fractal meta-resonator supercells and the cross-shape resonator derived from Ni–Mn–Sn alloy films. The Caley tree fractal meta-resonator supercells' host exhibits the desired broadband response, and the cross-shaped resonator exhibits the narrow-band response, while the aligned photoexcited Si layer modulates the resonance amplitude of the broadband absorber. Numerical results reveal that the absorption rate above 90% has a bandwidth of 1.129 THz from 1.950 THz to 3.079 THz. Notably, the function of broadband and narrowband absorption can be switched flexibly by applying a magnetic field. The narrowband mode shows an ultra-narrow bandwidth and a high-quality factor Q of ~25.8. The terahertz absorber proposed in this work has many advantages over the present advanced absorbers, including a thin 10.39-micron structure. Moreover, the absorption rate can be continuously tuned from 90% to 5% when the photoexcited Si transits from the insulator state to the metal state. In addition, the TMMA shows excellent performance in terms of polarization insensitivity and wide angle.

2. Structural Design and Method Simulation

The unit cell of the tunable TMMA that we designed is illustrated in Figure 1. It consists of Cayley tree metal resonators, a cross-shaped resonator, a photoexcited Si layer, a dielectric film, and a ground metal from top to bottom. The top layer of the unit cell consists of one, two, and three orders of fractal branches. To ensure polarization insensitivity, each branch is spaced apart with a 120° rotation angle, and the threefold rotational symmetry remains. The proposed structure uses the Cayley tree metal resonators to broaden the bandwidth. Simultaneously, the cross-shape resonator is added to the design for a narrowband. The metal of the Cayley tree metal resonators is Ni–Mn–Sn shape memory alloy films whose branches can be curved by applying an additional magnetic field, resulting in a switchable terahertz absorber from broadband to narrowband absorption. Ni–Mn–Sn is modeled as the lossy metal with a conductivity of 3.3 × 10^5 S/m [49]. The continuous silicon has simulated a dielectric with constant permittivity ε_{Si} = 11.9 [50] and pump-power-dependent conductivity σ_{Si}. The dielectric permittivity of sapphire is 9.4 without a loss for the dielectric film [51]. The ground metal is selected as copper with a conductivity of 5.96 × 10^7 S/m [52].

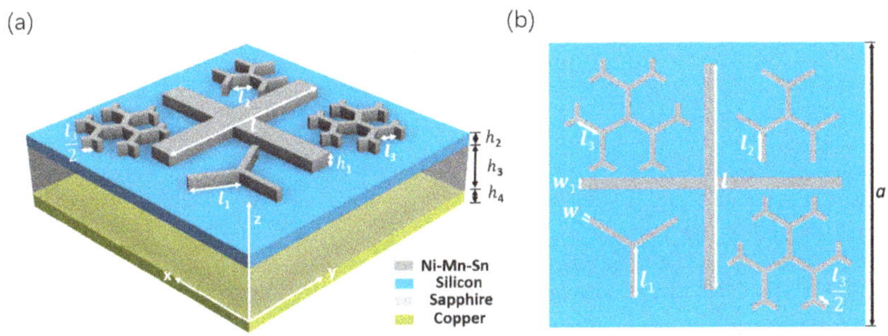

Figure 1. (**a**) 3D schematic of the designed switchable terahertz metamaterial unit cell. The unit cell is shown and includes the top metal (Ni–Mn–Sn) pattern over a bottom copper film separated by two dielectric spacers (Si and sapphire) with thicknesses h_1 = 0.2 μm, h_2 = 0.1 μm, h_3 = 10 μm, and h_4 = 0.1 μm, respectively. (**b**) The top view with detailed parameters. The arm length of the cross-shape is l = 40 μm, level 1 of the Caley tree fractal is l_1 = 8 μm, level 2 is l_2 = 4.4 μm, and level 3 is l_3 = 5.8 μm, whereas the outer arm length of level 3 is $l_3/2$. The arm widths are w_1 = 2 μm and w = 1 μm, respectively. The period of the unit cell is a = 59 μm.

The principle of Ni–Mn–Sn resonant film deformation is shown in Figure 2. When an external magnetic field is applied, the magnetic Ni–Mn–Sn shape memory alloy undergoes an anti-martensitic phase transformation from antiferromagnetic martensite to ferromagnetic austenite, where the direction of the austenitic magnetic induction axis eventually coincides with the direction of the external magnetic field, resulting in a large magnetic strain. When the external magnetic field is released, the alloy undergoes a martensitic phase transformation and returns from austenite to martensite. This process provides cyclically reversible large-output strains.

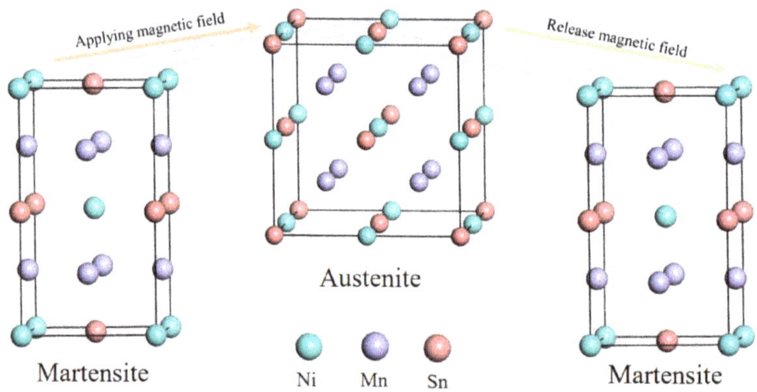

Figure 2. Schematic diagram of magnetic field induced Ni–Mn–Sn martensite phase transformation.

The materials present in nature are composed of atoms, and the electromagnetic response is determined with the dielectric constant and the magnetic permeability. Its response frequency and dielectric constant follow the Drude–Lorentz relationship:

$$\varepsilon_{eff}(\omega) = 1 - \frac{\omega_p^2}{\omega^2 - \omega_0^2 - i\gamma\omega} \tag{1}$$

The plasma frequency is determined with the following equation:

$$\omega_p = \frac{4\pi n e^2}{m^*} \tag{2}$$

where ω_0 is the resonant frequency, n is the carrier density, e is the charge of the electron, γ is the damping coefficient, and m^* is the carrier effective mass.

The absorber proposed in this paper is based on this principle, the designed sub-wavelength structure is atom-like, and the periodic macroscopic repetitive unit structure constitutes the whole metamaterial; therefore, its electromagnetic response is related to the subwavelength unit structure [49].

In this paper, Ni–Mn–Sn films are used as resonant structures because, in addition to their unique shape memory effect, they have a high virtual refractive index (R.I.) that can be sufficient to provide excellent absorption. The refractive index n, effective permittivity ε_{eff}, and magnetic permeability μ_{eff} of the metamaterial satisfy the following relationship:

$$n = \sqrt{\varepsilon_{eff}\mu_{eff}} \tag{3}$$

The martensitic phase transformation of Ni–Mn–Sn SMAs is accompanied by a change in shape as well as a change in material properties, both of which are actively modulated, affecting the electromagnetic response in terms of terahertz wave transmission characteristics and reflection properties. Among them, the dielectric constant is the direct evidence to characterize the change in material properties. Therefore, the mechanism of dynamic modulation of the electromagnetic response of metamaterials by shape memory alloys can be fully explained by using the variation of the dielectric constant. According to the Drude model, the permittivity in the terahertz band can be characterized with the following equation:

$$\varepsilon_m \approx -\sigma_{dc}/(\varepsilon_0\Gamma) + i\sigma_{dc}/(\varepsilon_0\omega) \tag{4}$$

where σ_{dc} is the direct current conductivity, Γ is the damping rate, and ε_0 is the vacuum dielectric. The outcome of THz–TDS is excellent, consistent with the simplified Drude model. In summary, both the real and imaginary parts of the permittivity contain direct current conductivity, so that the change in permittivity can be described by the change in conductivity. Accordingly, it is possible to directly characterize the reform in material properties throughout the martensitic phase transformation of SMAs in terms of electrical conductivity. In this simulation, we chose the Ni–Mn–Sn SMAs' conductivity before and after the phase change within the bounds of 3.3×10^5 S/m to 7.6×10^5 S/m.

Intending to analyze the performance of the designed absorber, simulations were performed using CST MICROWAVE STUDIO® (CST Computer Simulation Technology GmbH, Darmstadt, Germany). The unit cell condition should be defined at the x direction and y direction for the boundary condition, and the open (add space) was defined at the z direction for the boundary condition [53]. We defined the transverse electric (TE) polarization mode as the electric field parallel along the y-axis and the magnetic field parallel along the x-axis when the THz wave is normal incidence to the introduced absorber. Transverse magnetic (TM) polarization mode is the opposite. The unit cell's geometrical parameters of the absorber were optimized at h_1 = 0.1 um, h_2 = 0.1 um, h_3 = 10 um, h_4 = 0.1 um, l_1 = 8 um, l_2 = 4.4 um, l_3 = 5.8 um, w = 1 um, and w_1 = 2 um. However, in level 3 of the Caley tree fractal meta-resonator supercells, the outermost arm lengths were set to $l_3/2 = 2.9$ to prevent overlapping of neighboring structures. The metallic copper substrate leads to a transmittance of 0 (i.e., $S_{21} = 0$). The absorbance was calculated as $A(\omega) = 1 - R(\omega) - T(\omega) = 1 - |S_{11}|^2 - |S_{21}|^2 = 1 - |S_{11}|^2$.

In addition, the method of sample fabrication is important to the terahertz components. Recently, MEMS processing technology, such as the etching of silicon substrates, has made great progress [54] with thin-film deposition prior to gluing or by choosing a substrate with a thin film, such as silicon. Sapphire (epitaxial silicon on sapphire) and the desired structure are obtained by etching after hardening. During the etching process, the photoresist acts as protection, and the resulting structure has the same shape as the photoresist to ensure that the Caley tree structure in the proposed metamaterial is fixed to the silicon surface and remains stable without falling off. It is necessary to deposit a metal copper thin film on the Si layer with a method similar to the Refs. [55,56] thin-film structure, and then deposit the SMA layer on the copper thin film.

3. Results and Discussion

3.1. Broadband THz Absorber

The simulated absorption spectra of the Ni–Mn–Sn-based absorber are shown in Figure 3. Figure 3a–d shows the simulated absorption spectra of the first, second, and third iteration of the proposed Cayley tree resonating structures and the cross structure. For each structure, they cannot achieve broadband absorption independently. However, broadband absorption can be achieved by combing these four structures into one supercell, as shown in Figure 3e. It is observed that the absorption magnitude of the introduced metamaterial is more than 90% in the frequency range from 1.950 THz to 3.079 THz, suggesting that the proposed absorber has an excellent absorption performance. It is worth mentioning that the peak absorption magnitude is observed at multiple frequencies, i.e., 2.101, 2.524, and 2.692 to 3.019 THz. The near-unity absorption broadband spectrum caused by the mechanism of multiple resonance peaks originates from the different orders of the Caley tree fractal resonators [27]. Therefore, self-similar fractals are proposed as an ideal candidate for ultra-broadband absorbing devices due to the multiscale geometric features involved, which provide a degree of freedom for varying the resonance frequency.

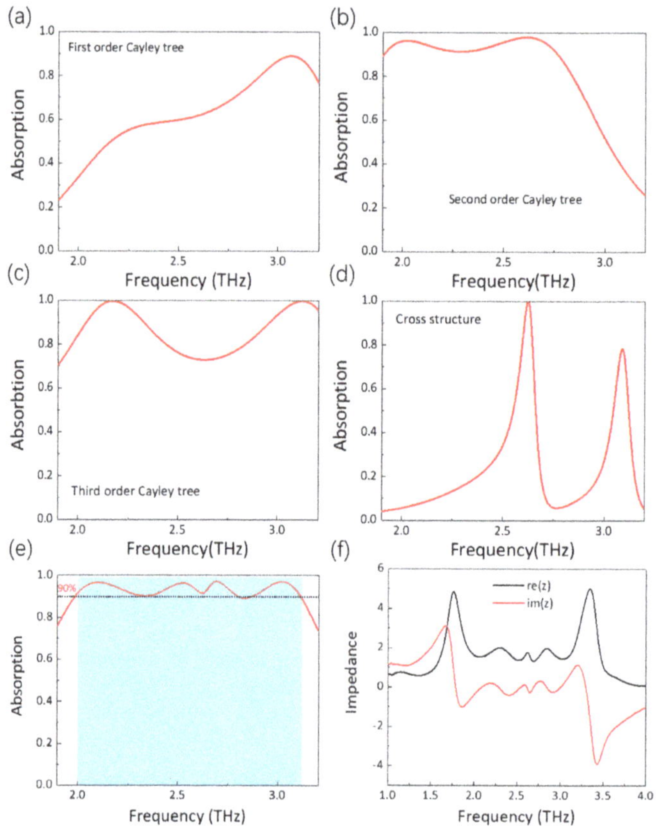

Figure 3. Absorption spectrum. (**a**) Absorption of first−order Cayley tree, (**b**) absorption of second-order Cayley tree, (**c**) absorption of third−order Cayley tree, (**d**) absorption of cross structure, (**e**) absorption spectrum of broadband terahertz metamaterial cell design. The 90% absorptivity and bandwidth are indicated by the short dotted line and shallow blue area, respectively. (**f**) Normalized effective impedance spectrum of the broadband terahertz metamaterial.

In addition, the relative bandwidth of the introduced absorber is also calculated through $f_c = 2(f_H - f_L)/(f_H + f_L)$, where f_H is the upper frequency, f_L is the lower frequency, and the relative bandwidth f_c is 45%. In order to reveal the working mechanism of the introduced single-layer Ni–Mn–Sn broadband absorber, we retrieved the corresponding effective impedance using the *S*-parameter extraction method [57,58]. The effective impedance can provide insight into understanding the optical response of the proposed metamaterial absorber. The relative impedance can be given by

$$Z_{eff}(\omega) = \pm\sqrt{\frac{(1+S_{11})^2 - S_{21}^2}{(1-S_{11})^2 - S_{21}^2}}, \tag{5}$$

where S_{11} and S_{21} are the complex reflection and transmission coefficients, respectively. Figure 3f shows the real and imaginary sections of the effectual impedance for the introduced metamaterial broadband absorber. The most significant finding to emerge from the following graph is that the real section is a short distance from 1, and the imaginary section is in the neighborhood of 0 from 1.950 THz to 3.079 THz. This phenomenon is consistent with the effectual impedances of the introduced metamaterial absorber matching better to the impedance of the free space, resulting in high absorption intensity. The large bandwidths are thought to be owed to the coupling of the adjacent resonances. Gottheim et al. proposed a detailed qualitative analysis to describe the mode degeneracies and origin of distinct absorption peaks in Cayley tree fractal geometry metamaterials [27]. In conclusion, the strong resonance frequency is mainly due to the first, second, and third fractal orders or the interaction between two levels.

3.2. Switchability between Broadband and Narrowband Absorption under the Magnetic Field

A significant bandwidth decline in the absorber can be driven by the structure change in the metamaterial Ni–Mn–Sn while applying an extraneous magnetic field. To understand the structural conversion of metamaterial in the shape of a Caley tree fractal based on Ni–Mn–Sn, we characterized the absorption at different curved angles (θ) of the Caley tree fractal resonators as schematically shown in Figure 4a. The simulated absorption spectrum was taken for the initial pristine structure of the absorber introduced, and the bendable part of the Caley tree fractal resonators was curved to 20°, 30°, 40°, and 50° (Figure 4b). The absorption patterns of the introduced device showed absorption peaks associated with the Caley tree fractal resonators. The narrowband forms at the interface between the Ni–Mn–Sn in a cross shape and the copper ground plane during the Ni–Mn–Sn Caley tree fractal metamaterials' deformation process. After the Ni–Mn–Sn Caley tree fractal metamaterials were curved to 50°, the broadband peaks disappeared. Only the narrow peak was still visible, and its absorption intensity was still maintained above 90%. This is consistent with the disappearance of the initial Ni–Mn–Sn Caley tree fractal metamaterial in a plane and the formation of Ni–Mn–Sn cross-shape metamaterial independent. It should be mentioned that, although the Caley tree part and the cross-shape part are both made of Ni–Mn–Sn FSMAs, their magnetic-induced martensitic transformation behavior can be controlled independently due to the slight difference in their alloy compositions and martensitic transformation temperatures [59,60]. When the Ni–Mn–Sn SMAS film in the proposed absorber is in the austenitic state, the absorber has the function of a broadband absorber, which is indispensable in thermophotovoltaic, photodetection, calorimetric, and mechanical resonance operations [25]. As the applied magnetic field increases, the martensitic phase transformation is gradually complete, and the Ni–Mn–Sn FSMAs undergo a shape change so that the absorber gradually changes from a broadband absorber to a narrowband absorber. Narrowband absorbers are widely used in food quality monitoring, disease diagnosis, biosensing, and other fields [40].

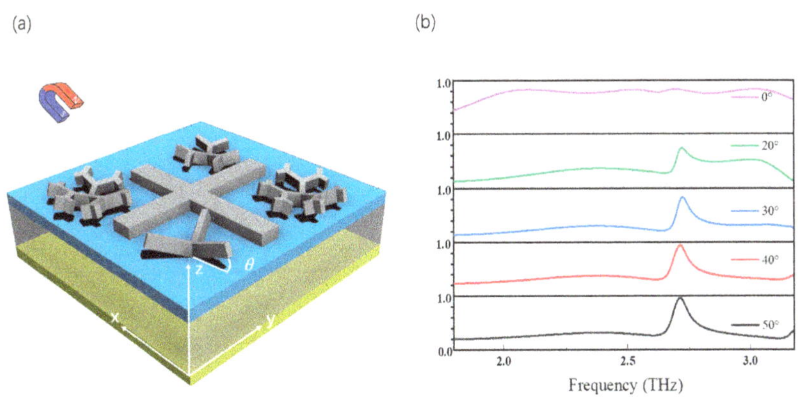

Figure 4. (**a**) Schematic drawing of Cayley tree fractal-shape metamaterials with different curved angles θ under an external magnetic field. (**b**) The calculated absorptances with different curved angles θ of Ni–Mn–Sn metamaterials in the shape of a Cayley tree fractal.

3.3. *Multifunction under the Optical Pump*

Further research will explore the influence of electromagnetic corresponding under the pump-power optical control on silicon film [61], as schematically shown in Figure 5a. When light shines on silicon, the light intensity increases, and the light carrier in extremely resistive silicon increases, leading to an increase in its electrical conductivity. A dynamic tunable intensity of absorption can be observed by changing the conductivity of photoexcited Si, as indicated in Figure 5b. The simulated absorption spectrum shows switching regimes from 1.950 THz to 3.079 THz frequencies with reasonably high contrast. From 1.950 THz to 3.079 THz, the metamaterial switches from a high-absorption state at small conductivity of Si to a low-absorption state at high conductivity of Si as the photoexcited Si changes its state from an insulating state to a metallic state. We observe that the absorption of the metamaterial structure decreases from 95% to 5% when the conductivity of Si increases from 1 S/m to 1 × 10^6 S/m [62,63]. Based on the Drude model, the photoexcited carrier mass results in an increase in the imaginary section of the dielectric constant, which leads to a bigger decrease when terahertz waves after the silicon wafer [64]. This phenomenon can be attributed to photoexcited Si due to the generated free carriers screening the confined resonant fields, which cannot store electromagnetic energy in the introduced device [65].

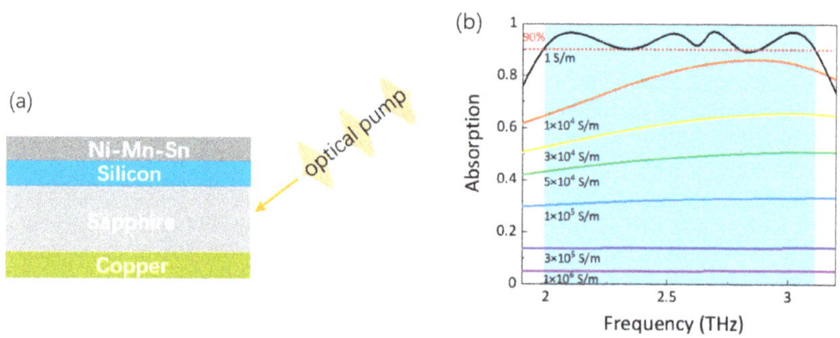

Figure 5. (**a**) Schematic of the terahertz metamaterial absorber side view under optical stimulus focus on the Si film. (**b**) The calculated absorptances with different conductivities of photoexcited Si. At a given resonance frequency, when 400 nm is incident on ion-irradiated Si, the photoconductivity of Si (σ) increases, thereby leading to the modulation of resonance strength.

As a result, the switching intensity of 90% is obtained in the switching range of 1.129 THz, and the corresponding switching contrast (SC), defined as SC = (Amax − Amin)/Amin, is calculated to be at least 1700% and can be improved by further increasing the conductivity of silicon, which is above six times higher than the previously reported results [66].

As the simulation results show, by using an applied magnetic field to drive the deformation of the shape memory alloy, we can achieve the conversion between wide/narrow absorbers and the function of absorber switching. In order to describe more clearly the advantages of this work, we compared it with some recently proposed tunable absorbers, as shown in Table 1. In addition, switching the total quality factor Q of the narrow-band absorber function with a maximum value Q of approximately 25.8 has potential applications in the sensing field. It is worth mentioning that the deformation of Ni–Mn–Sn SMAs by magnetic field modulation greatly reduces the complexity of the terahertz metamaterial design and facilitates device integration.

Table 1. Comparison of absorption performance between different absorbers.

References	Function	Working Bandwidth	Absorptance	Tunable Range	Material	Regulating Method
[30]	Switching	1.85–4.3 THz	>90%	4–100%	Vo_2	Temperature
[31]	Switching	4.29–5.52 THz	>90%	4–100%	Vo_2	Electric field
[32]	Switching	1–2.03 THz	>90%	0–99%	Graphene, Vo_2	Temperature, Electric field
[34]	Wide-narrow	3.57–8.45 THz	>50%	—	Vo_2	Temperature
[36]	Switching	1.2–2.67 THz	>90%	0–95%	MoS_2	Temperature
[16]	Wide-band	500–3000 nm	>90%	—	—	—
[20]	Dual-band	200–1000 nm	>90%	—	Au/Cr/3 layers of graphene/TMDs pair	—
This Work	Wide-narrow/Switching	1.950–3.08 THz	>90%	5–95%	Ni–Mn–Sn	Magnetic Fields

3.4. Incident Angle Sensitivity Characteristics

By rotating the direction incident to the terahertz wave placed in front of the terahertz metamaterial absorber, we further investigate and explore the incident angle sensitivity particularities of the broadband absorber. When the direction of the incident terahertz is perpendicular to the broadband absorber, the absorption intensities located at 1.950 THz to 3.079 THz are the maximum values, and the direction of the incident terahertz wave at this moment is set as the starting point (recorded as 0°). As shown in Figure 6a, when rotating the direction of the incident terahertz wave to 40°, the multi-resonance modes are both interaction modes, resulting in broadband and outstanding absorption. Analyzing the normalized curves of the absorption intensities of the TM mode shown in Figure 6b, the curves are approximately the same from 0° to 40°, which illustrates that the multi-resonance modes have the same interaction state. Therefore, it can be concluded that a stable broad-wavelength absorber can also be obtained even if the incident angle of the modes is 40°, which can be used to generate broadband absorption through the design introduced.

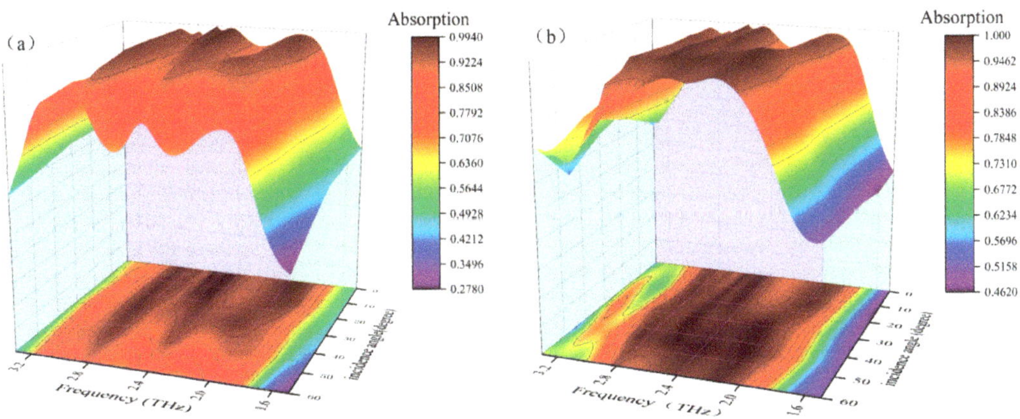

Figure 6. Numerically simulated incident angle dependent of the broadband terahertz metamaterial absorber: (**a**) TE mode and (**b**) TM mode.

3.5. Mechanism of Wide/Narrow Band Absorption Conversion under Magnetic Field

To comprehend the mechanism of forming narrowband absorption, the distributions of surface current and electric field are simulated and analyzed at the specific resonance frequency, as shown in Figure 7. Color maps and arrows denote the strength of the field and current, respectively. The electric field accumulates mainly in both ends of the vertical branches of the Ni–Mn–Sn cross shape when the incident wave is applied as a TE-polarization wave; for TM-polarized waves, the electric field concentrates mainly around the edges of the horizontal branches. The electric field distributions demonstrate that the incident terahertz wave excites typical electrical dipole resonances at the topside surface of the Ni–Mn–Sn cross patch along the *y* and *x* directions, respectively, for different polarization terahertz wave incidents (see Figure 7a,d).

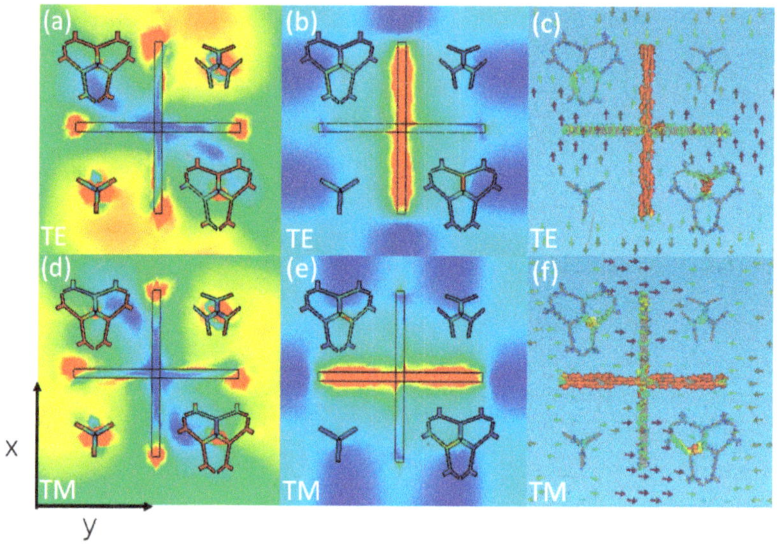

Figure 7. Distributions of (**a**,**d**) electric, (**b**,**e**) magnetic fields, and (**c**,**f**) surface currents at the resonant frequency of 2.71 THz for TE and TM-polarized THz wave in the inward propagation.

The contour map of the magnetic field distribution in the introduced absorber in the *x*-*y* plane is shown in Figure 7b,e. As revealed, the magnetic fields of the absorption band are all distributed over the cross shape of the narrowband superabsorber, which confirms the localized resonance characteristics of the absorption bands. However, the magnetic field distribution positions of the TE and TM polarizations are different. For TE-polarized waves, the magnetic field is mainly concentrated at the upright strip. Similar to the TM polarization wave, the magnetic field is mostly gathered at the horizontal strip. The arrows in Figure 7c,f represent the intensity and flowing directions of the surface current distributions on the top cross resonator and bottom ground plane. The surface current distribution of the TE polarization waves on the Ni–Mn–Sn SMA resonant films is principally around the *x*-axis direction, while the surface current on the metallic copper substrates is parallel to its inverse. Consequently, a magnetic polariton is considered [67], which can result in a large magnetic resonance and a deep, resonant dip in the reflection spectrum [68]. Meanwhile, for the TM-polarized waves, the anti-parallel surface currents give rise to the front resonator and back substrate along the horizontal direction, forming an equivalent current loop. The existence of excited electromagnetic resonance confines the incident terahertz waves in the absorber. It utilizes the ohmic loss in the metal, and the dielectric loss in the dielectric layer dissipates the incident terahertz waves in the absorber. Eventually, the perfect narrowband absorber was demonstrated with absorption, which is approximately 100%.

3.6. The Feasibility of Experimental Demonstration

Based on the simulation results above, we have systematically studied and demonstrated the possibility of our design. Here, the feasibility of the experimental demonstration is discussed and analyzed. Despite the fact that follow-up experiments are not executed in the existing research, a significant amount of associated works have proved that the fabrication of our proposed absorber can be realized with inexpensive cost and excellent precision. First, as an advanced actuator and sensor, shape memory alloys are used commercially in many fields, such as the magnetic actuators in automotive applications [69], the micro-actuators or artificial muscles in robotic applications [70], Boeing's variable geometry chevron in aerospace application [71], and so on. Moreover, with the development of SMA nano/micro-scale properties [72], shape memory alloy films with ultra-small sizes have been fabricated and used in micro-electro-mechanical systems (MEMS) [73]. The application of magnetic field-driven SMAs in terahertz devices has been successfully demonstrated experimentally [74]. In our work, the design of a terahertz absorber dependent on Ni–Mn–Sn FSMA films is simply attributable to the distinctive functional behaviors, for instance superelasticity and shape memory effects, and the magnetic field is easy and fast to apply. It is reasonable to expect that our proposed terahertz absorber will be successful in practical applications. However, Ni–Mn–Sn suffers from weak deformation recovery and large losses. In addition, the applied magnetic field to drive the magnetic Ni–Mn–Sn phase transition is large. This limits their practical application. These problems can be solved by adjusting the composition of the shape memory alloy and improving the fabrication process in our subsequent work.

4. Conclusions

In summary, we present a versatile, ultrathin, superior performance terahertz absorber with Ni–Mn–Sn FSMAs. The absorber enables the interconversion of broadband absorption and narrowband absorption. Numerical results show that the absorption rate is maintained above 90% in the 1.129 THz bandwidth from 1.950 THz to 3.079 THz. The magnetic field can modulate the absorber to switch between broadband and narrowband absorbers with the narrowband mode exhibiting an ultra-narrow bandwidth and a high-quality factor Q of ~25.8. In addition, the device can dynamically adjust the absorption rate from 90% to 5%. Notably, the absorber is insensitive to incidence angle, which allows for a wider range of applications. The terahertz absorber proposed in this work has many advantages over the present advanced absorbers, including the advantages of ultra-wide operating frequency, excellent absorption efficiency, and transient response. This work supports new ideas for the development of dynamically adjusted multifunctional terahertz functional components, and the proposed devices have potential applications in fields such as thermophotovoltaic energy conversion and sensors. The weak recovery of Ni–Mn–Sn deformation and the large driving magnetic field limit the practical application of the proposed absorber. However, these problems can be solved by tuning the composition of shape memory alloys and improving the fabrication process in our subsequent work.

Author Contributions: Conceptualization, K.Z. and C.T.; investigation, methodology, R.L. and X.W.; writing original draft preparation, R.L. and X.W.; writing—review and editing, K.Z. and C.T.; formal analysis, J.Z., X.T. and W.Z.; supervision, C.T. and K.Z.; resources and funding acquisition, J.Z., X.T., C.T. and K.Z. All authors have read and agreed to the published version of the manuscript.

Funding: This research is funded by the Heilongjiang Provincial Natural Science Foundation of China (Grant No. YQ2022112); the National Natural Science Foundation of China (Grant Nos. 51871083, 51971085, 52001101, 52271172).

Data Availability Statement: The data presented in this study are available upon request from the corresponding author.

Conflicts of Interest: The authors declare that they have no known competing financial interests or personal relationships that could have appeared to influence the work reported in this paper.

References

1. Lu, M.; Shen, J.; Li, N.; Zhang, Y.; Zhang, C.; Liang, L.; Xu, X. Detection and identification of illicit drugs using terahertz imaging. *J. Appl. Phys.* **2006**, *100*, 103104. [CrossRef]
2. Davies, A.G.; Burnett, A.D.; Fan, W.; Linfield, E.H.; Cunningham, J.E. Terahertz spectroscopy of explosives and drugs. *Mater. Today* **2008**, *11*, 18–26. [CrossRef]

3. Beard, M.C.; Turner, G.M.; Schmuttenmaer, C.A. Terahertz spectroscopy. *J. Phys. Chem. B* **2002**, *106*, 7146–7159. [CrossRef]
4. Taylor, Z.D.; Singh, R.S.; Bennett, D.B.; Tewari, P.; Kealey, C.P.; Bajwa, N.; Culjat, M.; Stojadinovic, A.; Lee, H.; Hubschman, J.-P.; et al. THz medical imaging: In vivo hydration sensing. *IEEE Trans. Terahertz Sci. Technol.* **2011**, *1*, 201–219. [CrossRef] [PubMed]
5. Federici, J.; Moeller, L. Review of terahertz and subterahertz wireless communications. *J. Appl. Phys.* **2010**, *107*, 6–323. [CrossRef]
6. Chen, H.T.; Padilla, W.J.; Zide, J.M.O.; Gossard, A.C.; Taylor, A.J.; Averitt, R.D. Active terahertz metamaterial devices. *Nature* **2006**, *444*, 597–600. [CrossRef] [PubMed]
7. Ju, L.; Geng, B.; Horng, J.; Girit, C.; Martin, M.; Hao, Z.; Bechtel, H.A.; Liang, X.; Zettl, A.; Shen, Y.R.; et al. Graphene plasmonics for tunable terahertz metamaterials. *Nat. Nanotechnol.* **2011**, *6*, 630–634. [CrossRef]
8. Linden, S.; Enkrich, C.; Wegener, M.; Zhou, J.; Koschny, T.; Soukoulis, C.M. Magnetic response of metamaterials at 100 terahertz. *Science* **2004**, *306*, 1351–1353. [CrossRef]
9. Bai, J.; Pang, Z.; Shen, P.; Chen, T.; Shen, W.; Wang, S.; Chang, S. A terahertz photo-thermoelectric detector based on metamaterial absorber. *Opt. Commun.* **2021**, *497*, 127184. [CrossRef]
10. Carranza, I.E.; Grant, J.P.; Gough, J.; Cumming, D. Terahertz metamaterial absorbers implemented in CMOS technology for imaging applications: Scaling to large format focal plane arrays. *IEEE J. Sel. Top. Quantum Electron.* **2016**, *23*, 1–8. [CrossRef]
11. Che, Z.; Zhang, G.; Ren, P.; Yue, J.; Li, Z.; Lun, Y.; Suo, J.; Zhu, J.; Zhang, Q.; Feng, Y. Narrow bandpass filter based on vanadium dioxide can be used for terahertz stealth. *J. Opt.* **2022**, *51*, 336–342. [CrossRef]
12. Landy, N.I.; Sajuyigbe, S.; Mock, J.J.; Smith, D.R.; Padilla, W.J. Perfect metamaterial absorber. *Phys. Rev. Lett.* **2008**, *100*, 207402. [CrossRef]
13. Shen, X.; Yang, Y.; Zang, Y.; Gu, J.; Han, J.; Zhang, W.; Jun Cui, T. Triple-band terahertz metamaterial absorber: Design, experiment, and physical interpretation. *Appl. Phys. Lett.* **2012**, *101*, 154102. [CrossRef]
14. Zhang, Y.; Feng, Y.; Zhu, B.; Zhao, J.; Jiang, T. Graphene based tunable metamaterial absorber and polarization modulation in terahertz frequency. *Opt. Express* **2014**, *22*, 22743–22752. [CrossRef] [PubMed]
15. Huang, L.; Chowdhury, D.R.; Ramani, S.; Reiten, M.T.; Luo, S.N.; Taylor, A.J.; Chen, H.T. Experimental demonstration of terahertz metamaterial absorbers with a broad and flat high absorption band. *Opt. Express* **2012**, *37*, 154–156. [CrossRef]
16. Agarwal, S.; Srivastava, G.; Prajapati, Y.K. Dual band Vis-IR absorber using bismuth based helical metamaterial surface. *Opt. Quantum Electron.* **2022**, *54*, 772. [CrossRef]
17. Islam, M.S.; Sultana, J.; Biabanifard, M.; Vafapour, Z.; Nine, M.J.; Dinovitser, A.; Cordeiro, C.M.B.; Ng, B.W.H.; Abbott, D. Tunable localized surface plasmon graphene metasurface for multiband superabsorption and terahertz sensing. *Carbon* **2020**, *158*, 559–567. [CrossRef]
18. Du, C.; Zhou, D.; Guo, H.H.; Pang, Y.Q.; Shi, H.Y.; Liu, W.F.; Su, J.Z.; Singh, C.; Trukhanov, S.; Trukhanov, A.; et al. An ultra-broadband terahertz metamaterial coherent absorber using multilayer electric ring resonator structures based on anti-reflection coating. *Nanoscale* **2020**, *12*, 9769–9775. [CrossRef]
19. Liu, Y.; Gu, S.; Luo, C.; Zhao, X. Ultra-thin broadband metamaterial absorber. *Appl. Phys. A* **2012**, *108*, 19–24. [CrossRef]
20. Agarwal, S.; Prajapati, Y.K. Design of broadband absorber using 2-D materials for thermo-photovoltaic cell application. *Opt. Commun.* **2018**, *413*, 39–43. [CrossRef]
21. Gu, S.; Barrett, J.P.; Hand, T.H.; Popa, B.I.; Cummer, S.A. A broadband low-reflection metamaterial absorber. *J. Appl. Phys.* **2010**, *108*, 064913. [CrossRef]
22. Hedayati, M.K.; Javaherirahim, M.; Mozooni, B.; Abdelaziz, R.; Tavassolizadeh, A.; Chakravadhanula, V.S.K.; Zaporojtchenko, V.; Strunkus, T.; Faupel, F.; Elbahri, M. Design of a perfect black absorber at visible frequencies using plasmonic metamaterials. *Adv. Mater.* **2011**, *23*, 5410–5414. [CrossRef]
23. Yu, P.; Besteiro, L.V.; Huang, Y.; Wu, J.; Fu, L.; Tan, H.H.; Jagadish, C.; Wiederrecht, G.P.; Govorov, A.O.; Wang, Z. Broadband metamaterial absorbers. *Adv. Opt. Mater.* **2019**, *7*, 1800995. [CrossRef]
24. Xie, T.; Chen, D.; Yang, H.; Xu, Y.; Zhang, Z.; Yang, J. Tunable broadband terahertz waveband absorbers based on fractal technology of graphene metamaterial. *Nanomaterials* **2021**, *11*, 269. [CrossRef]
25. Naveed, M.A.; Bilal, R.M.H.; Baqir, M.A.; Bashir, M.M.; Ali, M.M.; Rahim, A.A. Ultrawideband fractal metamaterial absorber made of nickel operating in the UV to IR spectrum. *Opt. Express* **2021**, *29*, 42911–42923. [CrossRef]
26. Bilal, R.M.H.; Naveed, M.A.; Baqir, M.A.; Ali, M.M.; Rahim, A.A. Design of a wideband terahertz metamaterial absorber based on Pythagorean-tree fractal geometry. *Opt. Mater. Express* **2020**, *10*, 3007–3020. [CrossRef]
27. Gottheim, S.; Zhang, H.; Govorov, A.O.; Halas, N.J. Fractal nanoparticle plasmonics: The Cayley tree. *ACS Nano* **2015**, *9*, 3284–3292. [CrossRef]
28. Zubair, A.; Zubair, M.; Danner, A.; Mehmood, M.Q. Engineering multimodal spectrum of Cayley tree fractal meta-resonator supercells for ultrabroadband terahertz light absorption. *Nanophotonics* **2020**, *9*, 633–644. [CrossRef]
29. Agarwal, S.; Prajapati, Y.K. Analysis of metamaterial-based absorber for thermo-photovoltaic cell applications. *IET Optoelectron.* **2017**, *11*, 208–212. [CrossRef]
30. Huang, J.; Li, J.; Yang, Y.; Li, J.; Li, J.; Zhang, Y.; Yao, J. Broadband terahertz absorber with a flexible, reconfigurable performance based on hybrid-patterned vanadium dioxide metasurfaces. *Opt. Express* **2020**, *28*, 17832–17840. [CrossRef]
31. Dao, R.N.; Kong, X.R.; Zhang, H.F.; Su, X.R. A tunable broadband terahertz metamaterial absorber based on the vanadium dioxide. *Optik* **2019**, *180*, 619–625. [CrossRef]

32. Zhou, R.; Jiang, T.; Peng, Z.; Li, Z.; Zhang, M.; Wang, S.; Li, L.; Liang, H.; Ruan, S.; Su, H. Tunable broadband terahertz absorber based on graphene metamaterials and VO2. *Opt. Mater.* **2021**, *114*, 110915. [CrossRef]
33. Agarwal, S.; Prajapati, Y.K. Broadband and polarization-insensitive helix metamaterial absorber using graphene for terahertz region. *Appl. Phys. A* **2016**, *122*, 561. [CrossRef]
34. Badri, S.H.; Gilarlue, M.M.; SaeidNahaei, S.; Kim, J.S. Narrowband-to-broadband switchable and polarization-insensitive terahertz metasurface absorber enabled by phase-change material. *J. Opt.* **2022**, *24*, 025101. [CrossRef]
35. Badri, S.H.; Soofi, H.; SaeidNahaei, S. Thermally reconfigurable extraordinary terahertz transmission using vanadium dioxide. *JOSA B* **2022**, *39*, 1614–1621. [CrossRef]
36. Zhong, Y.; Huang, Y.; Zhong, S.; Lin, T.; Luo, M.; Shen, Y.; Ding, J. Tunable terahertz broadband absorber based on MoS2 ring-cross array structure. *Opt. Mater.* **2021**, *114*, 110996. [CrossRef]
37. Liu, L.; Liu, W.; Song, Z. Ultra-broadband terahertz absorber based on a multilayer graphene metamaterial. *J. Appl. Phys.* **2020**, *128*, 093104. [CrossRef]
38. Wu, T.; Shao, Y.; Ma, S.; Wang, G.; Gao, Y. Broadband terahertz absorber with tunable frequency and bandwidth by using Dirac semimetal and strontium titanate. *Opt. Express* **2021**, *29*, 7713–7723. [CrossRef] [PubMed]
39. Liu, Y.; Huang, R.; Ouyang, Z. Numerical Investigation of Graphene and STO Based Tunable Terahertz Absorber with Switchable Bifunctionality of Broadband and Narrowband Absorption. *Nanomaterials* **2021**, *11*, 2044. [CrossRef]
40. Song, Z.; Chen, A.; Zhang, J. Terahertz switching between broadband absorption and narrowband absorption. *Opt. Express* **2020**, *28*, 2037–2044. [CrossRef] [PubMed]
41. Liu, W.; Song, Z. Terahertz absorption modulator with largely tunable bandwidth and intensity. *Carbon* **2021**, *174*, 617–624. [CrossRef]
42. Otsuka, K.; Ren, X. Physical metallurgy of Ti–Ni-based shape memory alloys. *Prog. Mater. Sci.* **2005**, *50*, 511–678. [CrossRef]
43. Cao, J.; Zhou, J.; Li, M.; Chen, J.; Zhang, Y.; Liu, X. Insightful understanding of three-phase interface behaviors in 1T-2H MoS_2/CFP electrode for hydrogen evolution improvement. *Chin. Chem. Lett.* **2022**, *33*, 3745–3751. [CrossRef]
44. Tan, C.; Liu, J.; Tian, X.; Zhu, J.; Zhang, K. Multifunctional and dynamically tunable terahertz metamaterials based on TiNi shape memory alloy films with a simple design. *Results Phys.* **2021**, *24*, 104165. [CrossRef]
45. Zhang, K.; Ma, T.; Liu, J.; Tian, X.; Zhu, J.; Tan, C. Dynamically tunable and polarization-insensitive dual-band terahertz metamaterial absorber based on TiNi shape memory alloy films. *Results Phys.* **2021**, *23*, 104001. [CrossRef]
46. Kainuma, R.; Imano, Y.; Ito, W.; Sutou, Y.; Morito, H.; Okamoto, S.; Kitakami, O.; Oikawa, K.; Fujita, A.; Kanomata, T.; et al. Magnetic-field-induced shape recovery by reverse phase transformation. *Nature* **2006**, *439*, 957–960. [CrossRef] [PubMed]
47. Thomas, M.; Heczko, O.; Buschbeck, J.; Lai, Y.W.; McCord, J.; Kaufmann, S.; Schultz, L.; Fähler, S. Stray-Field-Induced Actuation of Free-Standing Magnetic Shape-Memory Films. *Adv. Mater.* **2009**, *21*, 3708–3711. [CrossRef]
48. Zhang, K.; Tan, C.; Zhao, W.; Guo, E.; Tian, X. Computation-Guided Design of Ni–Mn–Sn Ferromagnetic Shape Memory Alloy with Giant Magnetocaloric Effect and Excellent Mechanical Properties and High Working Temperature via Multielement Doping. *ACS Appl. Mater. Interfaces* **2019**, *11*, 34827–34840. [CrossRef]
49. Wang, X.; Liu, X.; Zhang, K.; Liu, R.; Zhu, J.; Tian, X.; Tan, C. Magnetic-induced dual-function tunable THz polarization conversion metamaterial based on Ni-Mn-Sn shape memory alloy films. *Results Opt.* **2022**, *9*, 100274. [CrossRef]
50. Zheng, C.; Li, J.; Wang, S.; Li, J.; Li, M.; Zhao, H.; Hao, X.; Zang, H.; Zhang, Y.; Yao, J. Optically tunable all-silicon chiral metasurface in terahertz band. *Appl. Phys. Lett.* **2021**, *118*, 051101. [CrossRef]
51. Shabanpour, J. Programmable anisotropic digital metasurface for independent manipulation of dual-polarized THz waves based on a voltage-controlled phase transition of VO2 microwires. *J. Mater. Chem. C* **2020**, *8*, 7189–7199. [CrossRef]
52. Pan, W.; Yan, Y.; Ma, Y.; Shen, D. A terahertz metamaterial based on electromagnetically induced transparency effect and its sensing performance. *Opt. Commun.* **2019**, *431*, 115–119. [CrossRef]
53. Bisht, S.; Saini, S.; Prakash, V.; Nautiyal, B. Study the various feeding techniques of microstrip antenna using design and simulation using CST microwave studio. *Int. J. Emerg. Technol. Adv. Eng.* **2014**, *4*, 318–324.
54. Liu, A.Q.; Zhu, W.M.; Tsai, D.P.; Zheludev, N.I. Micromachined tunable metamaterials: A review. *J. Opt.* **2021**, *14*, 114009. [CrossRef]
55. Kim, H.; Melinger, J.S.; Khachatrian, A.; Charipar, N.A.; Auyeung, R.C.Y.; Piqué, A. Fabrication of terahertz metamaterials by laser printing. *Opt. Lett.* **2010**, *35*, 4039–4041. [CrossRef]
56. Walther, M.; Ortner, A.; Meier, H.; Löffelmann, U.; Smith, P.J.; Korvink, J.G. Terahertz metamaterials fabricated by inkjet printing. *Appl. Phys. Lett.* **2009**, *95*, 251107. [CrossRef]
57. Smith, D.R.; Vier, D.C.; Koschny, T.; Soukoulis, C.M. Electromagnetic parameter retrieval from inhomogeneous metamaterials. *Phys. Rev. E* **2005**, *71*, 036617. [CrossRef] [PubMed]
58. Smith, D.R.; Schultz, S.; Markoš, P.; Soukoulis, C.M. Determination of effective permittivity and permeability of metamaterials from reflection and transmission coefficients. *Phys. Rev. B* **2002**, *65*, 195104. [CrossRef]
59. Auge, A.; Teichert, N.; Meinert, M.; Reiss, G.; Hütten, A.; Yüzüak, E.; Dincer, I.; Elerman, Y.; Ennen, I.; Schattschneider, P. Thickness dependence of the martensitic transformation, magnetism, and magnetoresistance in epitaxial Ni-Mn-Sn ultrathin films. *Phys. Rev. B* **2012**, *85*, 214118. [CrossRef]
60. Krenke, T.; Acet, M.; Wassermann, E.F.; Moya, X.; Mañosa, L.; Planes, A. Martensitic transitions and the nature of ferromagnetism in the austenitic and martensitic states of Ni-Mn-Sn alloys. *Phys. Rev. B* **2005**, *72*, 014412. [CrossRef]

61. Rao, Y.; Pan, L.; Ouyang, C.; Xu, Q.; Liu, L.; Li, Y.; Zhang, W. Asymmetric transmission of linearly polarized waves based on Mie resonance in all-dielectric terahertz metamaterials. *Opt. Express* **2020**, *28*, 29855–29864. [CrossRef]
62. Chen, H.T. Interference theory of metamaterial perfect absorbers. *Opt. Express* **2012**, *20*, 7165–7172. [CrossRef] [PubMed]
63. Gu, J.; Singh, R.; Liu, X.; Zhang, X.; Ma, Y.; Zhang, S.; Zhang, W. Active control of electromagnetically induced transparency analogue in terahertz metamaterials. *Nat. Commun.* **2012**, *3*, 1151. [CrossRef]
64. Li, J.S.; Li, X.J. Switchable tri-function terahertz metasurface based on polarization vanadium dioxide and photosensitive silicon. *Opt. Express* **2022**, *30*, 12823. [CrossRef] [PubMed]
65. Bing, P.; Guo, X.; Wang, H.; Li, Z.; Yao, J. Characteristic analysis of a photoexcited tunable metamaterial absorber for terahertz waves. *J. Opt.* **2019**, *48*, 179–183. [CrossRef]
66. Pitchappa, P.; Kumar, A.; Liang, H.; Prakash, S.; Wang, N.; Bettiol, A.A.; Venkatesan, T.; Lee, C.; Singh, R. Frequency-Agile Temporal Terahertz Metamaterials. *Adv. Opt. Mater.* **2020**, *8*, 2000101. [CrossRef]
67. Ou, J.Y.; Plum, E.; Zhang, J.; Zheludev, N.I. An electromechanically reconfigurable plasmonic metamaterial operating in the near-infrared. *Nat. Nanotechnol.* **2013**, *8*, 252–255. [CrossRef]
68. Butera, F. Shape memory actuators for automotive applications. *Adv. Mater. Process.* **2008**, *166*, 37.
69. Jani, J.M.; Leary, M.; Subic, A.; Gibson, M.A. A review of shape memory alloy research, applications and opportunities. *Mater. Des.* **2014**, *56*, 1078–1113. [CrossRef]
70. Strelec, J.K.; Lagoudas, D.C.; Khan, M.A.; Yen, J. Design and implementation of a shape memory alloy actuated reconfigurable airfoil. *J. Intell. Mater. Syst. Struct.* **2003**, *14*, 257–273. [CrossRef]
71. Karami, M.; Chen, X. Nanomechanics of shape memory alloys. *Mater. Today Adv.* **2021**, *10*, 100141. [CrossRef]
72. Gupta, B.; Pandey, S.; Nahata, A.; Zhang, T.; Nahata, A. Bistable physical geometries for terahertz plasmonic structures using shape memory alloys. *Adv. Opt. Mater.* **2017**, *5*, 1601008. [CrossRef]
73. Zhou, H.; Zhang, T.; Guruswamy, S.; Nahata, A. An electrically tunable terahertz plasmonic device based on shape memory alloys and liquid metals. *Adv. Opt. Mater.* **2018**, *6*, 1700684. [CrossRef]
74. Valente, J.; Ou, J.Y.; Plum, E.; Youngs, I.J.; Zheludev, N.I. Reconfiguring photonic metamaterials with currents and magnetic fields. *Appl. Phys. Lett.* **2015**, *106*, 111905. [CrossRef]

metals

MDPI

Article

Enhanced Fatigue Resistance of Nanocrystalline $Ni_{50.8}Ti_{49.2}$ Wires by Mechanical Training

Peng Chen [1,2], Xiaorong Cai [1,2,*], Na Min [3], Yunfan Liu [4], Zhengxiong Wang [1], Mingjiang Jin [1,*] and Xuejun Jin [1,2]

[1] Institute of Phase Transformation and Complex Microstructure, School of Materials Science and Engineering, Shanghai Jiao Tong University, Shanghai 200240, China
[2] Institute of Medical Robotics, Shanghai Jiao Tong University, Shanghai 200240, China
[3] Key Laboratory for Microstructures, Shanghai University, Shanghai 200444, China
[4] Institute of Forming Technology and Equipment, School of Materials Science and Engineering, Shanghai Jiao Tong University, Shanghai 200030, China
* Correspondence: cxr1110@sjtu.edu.cn (X.C.); jinmj@sjtu.edu.cn (M.J.)

Abstract: In this paper, the fatigue resistance of superelastic NiTi shape memory alloy (SMA) wires was improved by combining mechanical training and nanocrystallization. Fatigue tests were performed after mechanical training with a peak stress of 600 MPa for 60 cycles of nanocrystalline (NC) NiTi wires, and the associated microscopic mechanism was investigated by using transmission electron microscopy (TEM) and transmission Kikuchi diffraction (TKD). The results showed that stress-controlled training effectively improved the functional stability (the accumulated residual strain decreased by 83.8% in the first 5000 cycles) of NC NiTi SMA wires, as well as increased the average structural fatigue life by 187.4% (from 4538 cycles to 13,040 cycles). TEM observations and TKD results revealed that training-induced dislocations resulted in lattice rotation and preferential grain orientation. The finite element method (FEM) simulation results indicated that the training-induced preferential grain orientation tended to decrease the local stress concentration and strain energy density. Combined with fractography analysis, the uniform deformation caused by mechanical training changed the crack growth mode from multi-regional propagation to single-regional propagation, improving the structural fatigue life.

Keywords: shape memory alloys; nanocrystalline materials; fatigue; mechanical training; transmission Kikuchi diffraction

1. Introduction

NiTi shape-memory alloys (SMAs) are frequently employed in biomedical devices [1,2], aerospace engineering [3], and automotive applications [4] due to their distinctive superelasticity. In practice, thousands of cyclic martensitic phase transformations are generally performed to NiTi SMAs. Fatigue failure is one of the main remaining challenges in the application of SMAs. As previously reported, NiTi SMA wires subjected to thermomechanical cyclic loading within a certain strain amplitude (approximately 3.5%) tend to fail at thousands of cycles [5,6], severely restricting the commercial applications of NiTi SMA wires.

In contrast to that of traditional metals, fatigue of NiTi SMA involves functional and structural fatigue [7]. The functional degradation mainly refers to the gradual accumulation of residual strain (ε_{res}) and the decrease in critical stress triggering stress-induced martensitic transformation (σ_{SIM}) with increasing cycles. The functional degradation is primarily associated with the formation of dislocations near the austenite–martensite interfaces to coordinate the inconsistent deformation [8]. The reduction in grain size down to the nanometer scale is a practical and useful method for enhancing functional stability [9,10] due to the increased lattice compatibility [11]. Structural fatigue is governed by crack nucleation and propagation [12,13]. Compared to conventional structural engineering materials, stress-induced martensitic transformation occurs at the crack tips, along with plastic deformation for NiTi SMAs [14–17]. In this case, the increase in grain size results in enlarged plastic zones and phase transformation zones, which leads to a higher energy dissipation ahead of the crack tips [18,19]. Therefore, the fatigue crack growth rate can be lowered by increasing the grain size due to the subsequent increased resistance to crack propagation. However, previous experiments show that a longer structural fatigue life can be obtained by decreasing the grain size [20], indicating that the reduction of grain size may be benefit to delay crack nucleation. As mentioned above, it is difficult to restrain both crack nucleation and propagation at the same time by controlling grain size alone. Although the structural fatigue life can be increased as the grain size is decreased to the nanoscale range, great opportunity for further improvement remains.

Mechanical training is also a key method to improve the fatigue resistance of NiTi alloys [21–23]. During training, dislocations are generated in NiTi alloys to coordinate the deformation. Pre-introduced dislocations in the training process are considered to suppress the accumulation of defects in the subsequent cyclic loading fatigue process, resulting in a more stable stress–strain response [24,25]. Additionally, training is also beneficial to improve the structural fatigue life of NiTi SMA wires. Zhang et al. [6] reported that the dislocation-induced local residual stress field caused by stress-controlled training at 700 MPa for 20 cycles significantly reduces the macroscopic stress needed for martensite transformation in the subsequent strain-controlled fatigue test and improves the structural fatigue life by 10 times up to 18,242 cycles under a maximum strain of 2.5%. Nevertheless, the detailed microscopic mechanism of the effect of training on structural fatigue is still unclear.

In previous studies, the mechanical training was mainly performed on coarse-grain NiTi alloys. However, nanocrystalline NiTi wires generally show a better functional stability and a higher structural fatigue life [9,10,20]. In this paper, the mechanical training is applied to nanocrystalline NiTi wires to enhance the fatigue resistance, which refers to both the reduction of the functional fatigue and improvement of the structural fatigue life. Transmission electron microscopy (TEM) and transmission Kikuchi diffraction (TKD) were used to characterize the evolution of microstructures after training, including dislocation structures, grain orientations, and kernel average misorientation (KAM), as well as the distribution of grain boundary misorientation. Combined with finite element simulations, the present study revealed that the fatigue performance of nanocrystalline NiTi SMA wires can be further improved by mechanical training and proposes a comprehensive explanation for its microscopic mechanism.

2. Materials and Methods

Commercial polycrystalline superelastic $Ni_{50.8}Ti_{49.2}$ SMA wires of 0.15 mm in diameter used in this study were obtained from PeierTech, Jiangyin, China. After heat treatment at temperature of 733 K for 5 min, followed by cooling in air, the NiTi wires were mainly composed of the B2 phase, with an average grain size of 84 nm (Figure 1). The NiTi wires tested in this study were divided into two groups. One group was subjected to stress-control training, following by the strain-control fatigue test. The other group was directly subjected to the strain-control fatigue test for comparative analysis. The former is referred to as the "trained" sample, and the latter is the "as-received" sample. The "trained" sample underwent 60 stress-control cycles, with a stress rate of 20 MPa/s (Figure S1 in Supplementary material shows the Selection principle of training system). During each cycle, the sample was loaded from 0 to 600 MPa and then relaxed from 600 MPa back to the stress-free condition. The strain-control fatigue tests on superelastic NiTi wires for both the as-received and trained samples were performed by controlling the applied displacement, and the maximum tensile strain used in the cyclic tests was set to 5%. All the sample wires were cycled until failure, and the strain rate used for the superelastic fatigue tests was chosen to be 1×10^{-2}/s. The length of nanocrystalline (NC) NiTi wires used for tests in the present study was approximately 15 mm for each sample. The training process and fatigue tests were performed on an ElectroForce DMA 3200 fatigue testing machine (TA Instruments, New Castle, DE, USA) at room temperature (296 ± 1 K). The experimental setup is shown in Figure 2. Two pairs of aided grips were adopted to hold the SMA wire in place to avoid breakage at the contact point under low cycle fatigue loading. Each test was repeated at least three times. The evolution of functional characteristics, such as ε_{res}, σ_{SIM}, total strain (ε_{total}), and pseudoelastic strain (ε_{pse}), was quantitative analyzed from the measured stress–strain responses.

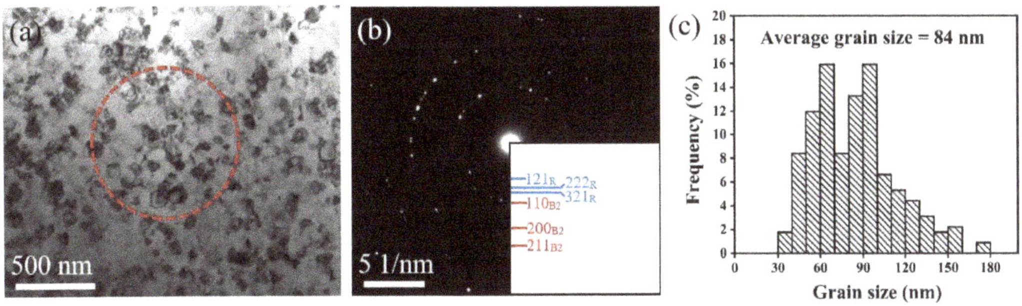

Figure 1. (**a**–**c**) show the bright-field TEM image, the SAED of the circled region in (**a**) and grain size distribution of the as-received NC NiTi sample, respectively.

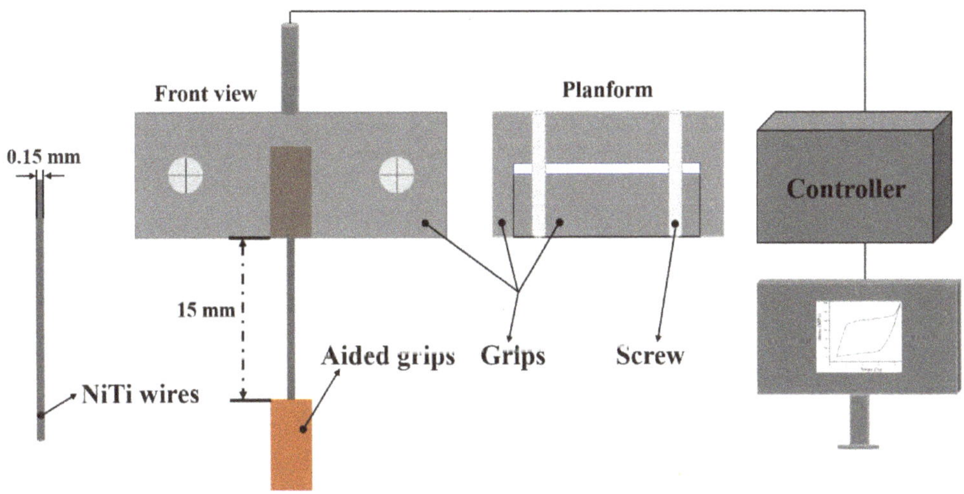

Figure 2. The geometrical representations of the real samples and the fatigue analysis apparatus.

A JEM ARM-200F (JEOL, Tokyo, Japan) TEM instrument was employed to determine the grain sizes and microstructures at 200 kV. A precision ion thinner (PIPS 695, Gatan, Pleasanton, CA, USA) was used to fabricate the TEM samples, and the wires were manually ground to a 40 µm thickness using 800-grit sandpaper. The grain orientations and grain boundaries were characterized by the TKD technique, with a significantly improved spatial resolution in the range of 2–10 nm [26,27]. The ion thinned samples were used for the TKD scans, and a 500 × 500 nm^2 area was scanned with a step size of 6 nm for each sample using a scanning electron microscope (SEM, TESCAN MIRA3, Brno, Czech Republic) equipped with an electron back-scatter diffraction (EBSD) detector. Fractography analysis was also performed using scanning electron microscopy in the secondary electron mode.

3. Results

3.1. Mechanical Characterization

The uniaxial strain-controlled cyclic fatigue tests were performed with a maximum strain of 5% for the as-received and trained samples, until failure occurred. The superelastic stress–strain responses and corresponding analysis are given in Figures 3 and 4. The associated functional characteristics, i.e., ε_{res}, ε_{pse}, and σ_{SIM}, are determined from the stress–strain curves, as schematically shown in Figure 3a. The accumulation of ε_{res} at cycle N is defined as the initial strain in the cycle N+1, as shown in Figure 3a.

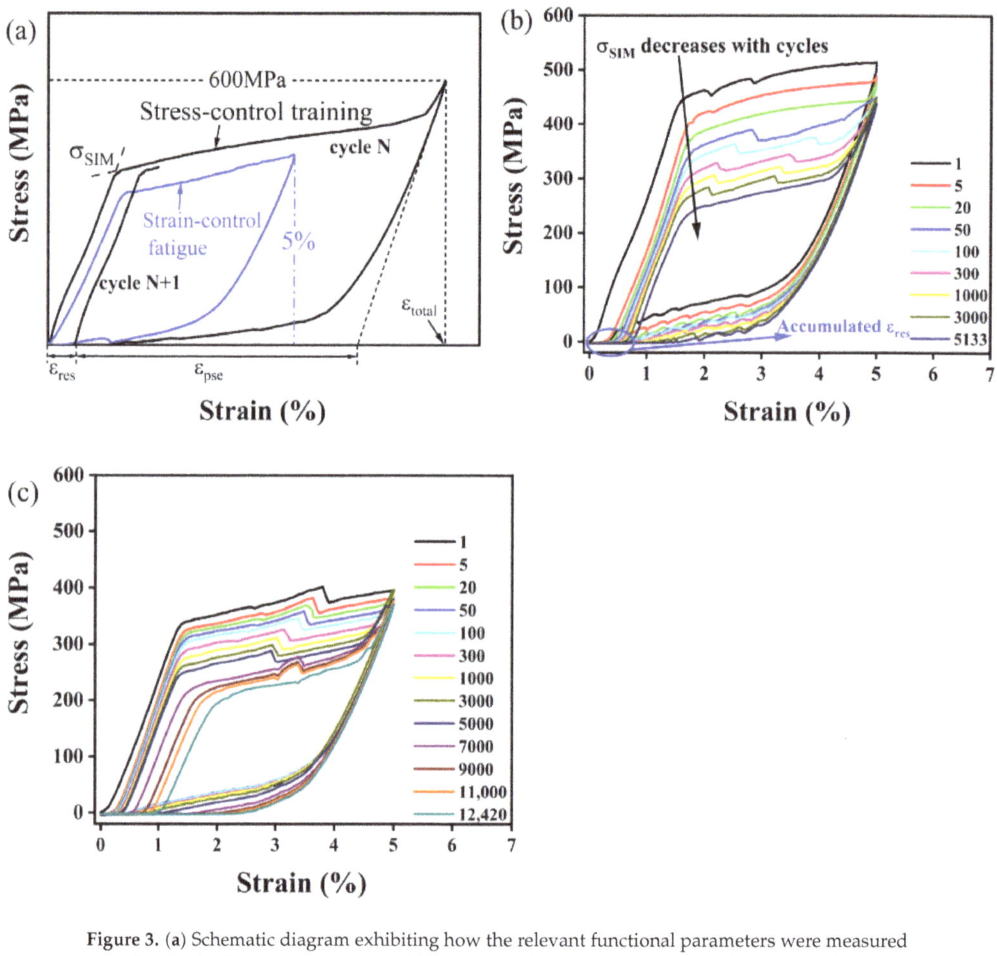

Figure 3. (**a**) Schematic diagram exhibiting how the relevant functional parameters were measured from the obtained stress–strain curves. The macroscopic stress–strain curves of nanocrystalline (NC) NiTi wires during fatigue tests until fracture: (**b**) as-received sample, and (**c**) trained sample.

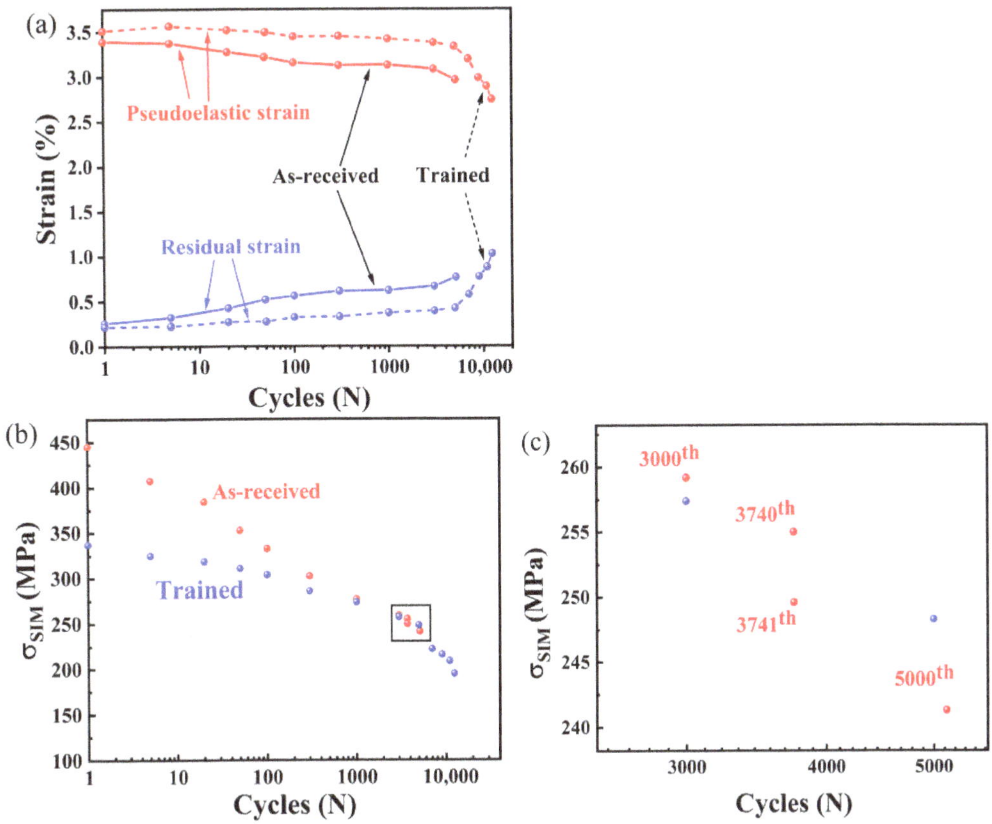

Figure 4. (**a**) indicates the evolution of pseudoelastic and residual strains, and (**b**) indicates the evolution of σ_{SIM} of the as-received and trained samples during fatigue tests. (**c**) is an enlarged view of the framed area in (**b**).

The evolution of residual and pseudoelastic strains with continuing cycles were calculated from the stress–strain curves in Figure 3b,c and shown in Figure 4a. According to Figure 4a, the residual strain increased during the fatigue tests, whereas the pseudoelastic strain decreased for both the as-received and trained samples. Compared to the as-received sample, the accumulated residual strain of the trained sample decreased by 83.8% in the first 5000 cycles, demonstrating that training suppressed dislocation accumulation during the fatigue tests. Figure 4b shows that σ_{SIM} decreased with repeated cycles for both samples. Considering ε_{res}, ε_{pse}, and σ_{SIM}, the functional degradation for the trained sample was much slower compared to that of the as-received sample.

Notably, the evolution of residual strain, pseudoelastic strain, and σ_{SIM} all exhibited two-stage evolutions. Figure 4 is plotted with a logarithmic scale, and each curve had two different slopes. Figure 4c is an enlarged figure of the black square indicated in Figure 4b. According to Figure 4a–c, residual strain, pseudoelastic strain, and σ_{SIM} evolved gradually during the first 3000 and 5000 cycles for the as-received and trained samples, respectively, followed by a sharp change during the subsequent cycles. Compared with those of the as-received sample, the second stage continues over cycles. The two-stage evolution phenomenon was probably evidence of a change in the failure mode from functional degradation to structural fatigue. The difference in the rate and duration of the evolution of functional parameters between the as-received and trained samples could be because the rate of crack propagation in these two samples was different. The details will be discussed in Section 4.

In addition to the improvement in functional stability, training also significantly increased the structural fatigue life of NiTi wires. In this case, a 187.4% improvement in average structural fatigue life after training was observed. Figure 5 presents a comparison of the number of cycles to failure for the superelastic NiTi SMA wires under pull–pull cycles described in this work and previous studies [5,6,28–31]. The structural fatigue life usually decreases with an increase in the maximum strain applied during the strain-control fatigue tests. The number of cycles to failure for the two as-received samples was 5133 and 3943, which was in line with the trend presented by the data in the literature. However, the fatigue life of the two trained samples reached 13,660 and 12,420 under the maximum strain of 5%, which was much higher than that reported in the literature. Thus, reasonable training was an effective means to improve fatigue resistance. The trained NiTi wires with outstanding fatigue resistance in the present study are a promising candidate material for the execution component with a lager applied strain.

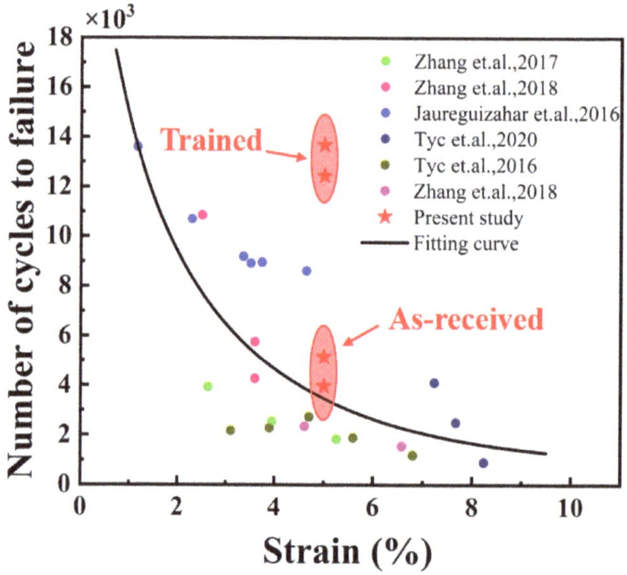

Figure 5. Comparison of structural fatigue life for NiTi SMA wires reported in previous research and this work [5,6,28–31].

To investigate the underlying mechanism of training in the enhancement of fatigue resistance, the macroscopic mechanical responses during the training and evolution of the microstructure were analyzed. Figure 6 shows the evolution of total strain, residual strain, pseudoelastic strain, and σ_{SIM} that were evaluated from the stress–strain responses during training. The measurement method of the functional parameters is schematically shown in Figure 3a. According to Figure 6, residual strain increased during the first 40 cycles and then reached a stable stage, followed by relatively slow growth. After 60 cycles, the accumulated residual strain reached 4.16%. The dislocations that accompany the phase transformation are usually responsible for the accumulation of residual strain [32]. σ_{SIM} was 429.4 MPa for the first cycle, and decreased rapidly at first, and then gradually stabilized at approximately 300 MPa. The decrease in σ_{SIM} possibly resulted from the internal stress field caused by training-induced dislocations.

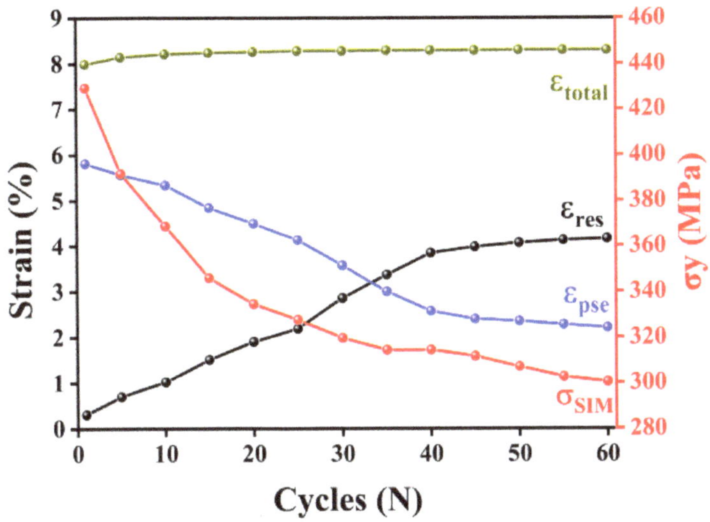

Figure 6. Evolution of critical stress to trigger stress-induced martensitic transformation (σ_{SIM}), total strain (ε_{total}), residual strain (ε_{res}), and pseudoelastic strain (ε_{pse}) during 60 cycles of training.

3.2. Microstructure Analysis

Figure 7a,b show the grain orientation maps for the as-received and the trained samples in the drawing direction (DD), transverse direction (TD), and normal direction (ND). Figure 7c,d show the corresponding inverse pole figures (IPFs) of these samples. The sample coordinate system used in the experiment is indicated in Figure 7a. The orientation of each grain is represented in IPF colors, and the black regions are undetected points according to TKD due to the high strain concentration. As shown in Figure 7a, the as-received sample had a weak <111> | | DD texture with the maximum orientation intensity of 5.3 (Figure 7c). There was no obvious preferred orientation in the TD and ND. After the training process, the sample had a strong texture in the <111> orientation parallel to the DD, and the maximum orientation intensity increased to 11.1 (Figure 7d). In the TD and ND, the orientations of the grains were also more concentrated compared to those of the as-received sample. Based on the TKD results, we propose that the lattice rotation occurred during the training process, thus resulting in a strong <111> | | DD texture. The Euler angles of each grain shown in Figure 7 were used to evaluate the amount of rotation due to training, and the calculated average rotation angle was 18° (Section 2, Supplementary Materials). The method of rotation angle computation as shown in ref. [33,34].

The comparison of the grain boundaries and the misorientations between the as-received and trained samples is shown in Figure 8a,b. Figure 8a shows that in the as-received sample, high-angle grain boundaries (HAGBs, >10°, marked by blue lines) were dominant, accounting for 70.6% of the boundaries, whereas the fraction of low-angle grain boundaries (LAGBs, 2–10°, marked by red lines) was only 29.4%. After training, the fraction of LAGBs increased to 59.6%. In some cases, several adjacent grains could not be distinguished in the IPF maps (the interfaces are marked by the white ovals in Figure 7b). Figure 8c,d presents the distributions of the KAM of the as-received and trained samples, respectively. Higher local misorientation regions were mostly distributed at the LAGBs, indicating that training-induced dislocations were associated with the formation of LAGBs. Figure 8e shows that the distribution of the local misorientations and average KAM values of the as-received and trained samples were 0.374 and 0.549, respectively. This result indicated that the geometrically necessary dislocations (GNDs) density increased by 46.8% after training because the GND density is proportional to the KAM value [35,36].

The significantly increased volume fraction of LAGBs with a high density of GNDs for the trained sample provided more evidence that lattice rotation occurred during training.

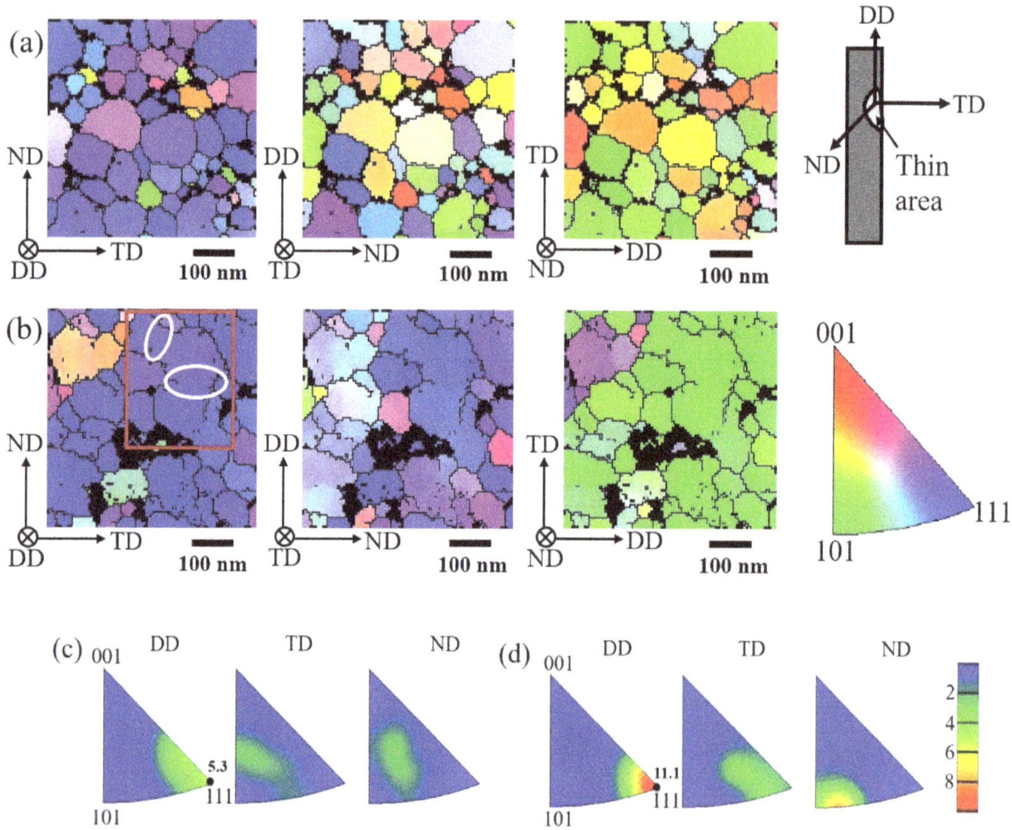

Figure 7. (**a**,**b**) The inverse pole figure (IPF) maps of the as-received and trained sample, respectively; (**c**,**d**) IPFs of the as-received and trained sample, respectively.

Figures 9–11 show microstructural observation by TEM of the as-received and trained samples. According to Figure 9, we can see that there were few dislocations in the as-received sample. Geometrical phase analysis (GPA) was used to analyze the relevant atomic strain distribution (Figures 9d,e and 10h–k). Figure 9d,e shows that the atomic strain distribution was uniform, with a relatively low level in the as-received sample. Figure 10a is the bright-field image of the trained sample. Figure 10a shows the existence of dislocation walls, which are marked by blue arrows, inside the grains in the trained sample. The high-resolution transmission electron microscope (HRTEM) image of the blue rectangle area in Figure 10a revealed that the dislocation wall divided the grain into two parts (Figure 10b). Figure 10c–e shows the fast Fourier transformation (FFT) patterns of the green, yellow, and red square domain in Figure 10b. The indexed FFT pattern sketch shown in Figure 10f reveals that there was a misorientation between the lattice planes on the two sides of the dislocation wall, indicating that the lattice rotation in a single grain was uneven. Figure 10g shows the inverse FFT pattern (IFFT) of area-1. Only a few dislocations remained in area-1, as identified in the IFFT image. Figure 10h,i and Figure 10j,k present the GPA results of the matrix (area-1) and the dislocation wall region (area-3), respectively. The atomic strain distribution was sparse in area-1, whereas this distribution was dense in area-3 in both the xx and yy directions. This result indicated that a higher degree of lattice distortion occurred in the region closer to the dislocation walls, corresponding to the higher density dislocations in area-3, as shown in Figure 10l.

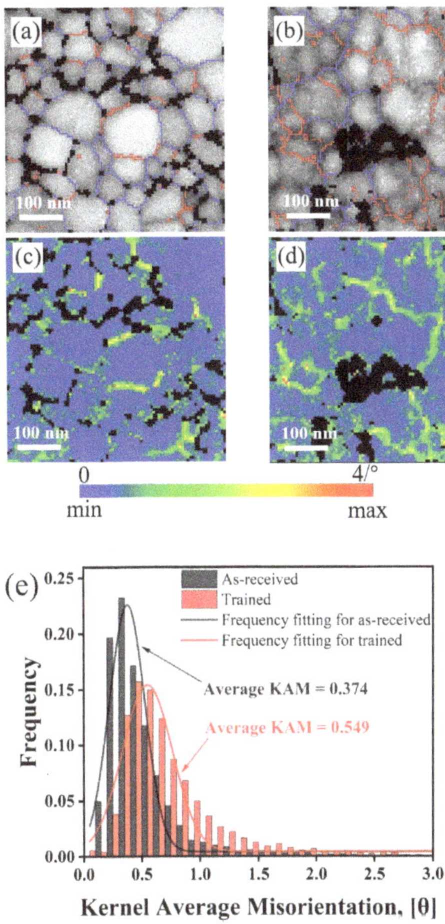

Figure 8. (**a**,**c**) show grain boundaries and kernel average misorientation (KAM) maps for the as-received sample, respectively. (**b**,**d**) are those for the trained sample. For (**a**,**b**), low-angle grain boundaries (LAGBs) are in red (2° < β < 10°), and high-angle grain boundaries (HAGBs) are in blue (β > 10°). (**e**) shows the distribution of KAM values for both samples.

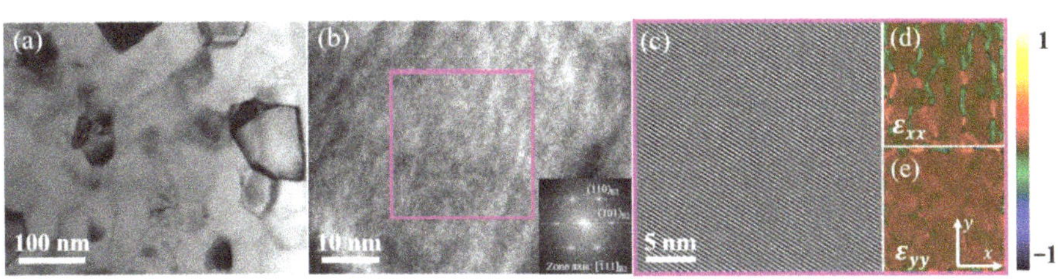

Figure 9. TEM observation of the as-received sample: (**a**) bright-field images; (**b**) high-resolution transmission electron microscope (HRTEM) images; (**c**–**e**) show the inverse fast Fourier transformation (IFFT) images and the geometric phase analysis (GPA) images of the pink square domain in (**b**).

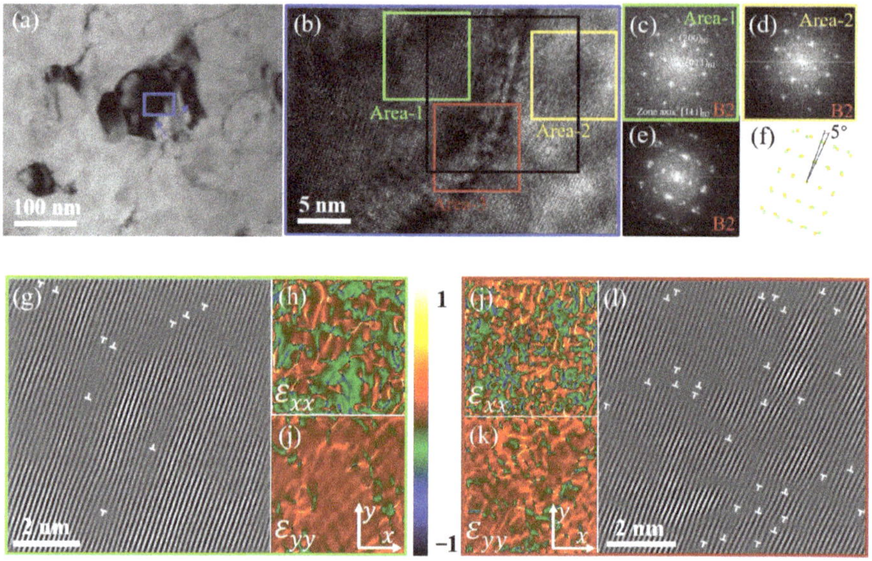

Figure 10. TEM observation of the trained sample: (**a**) bright-field image; (**b**) HRTEM image; (**c–e**) show the fast Fourier transformation (FFT) patterns; (**f**) is an indexed FFT pattern sketch of (**e**); (**g**,**l**) show the IFFT of (**c**,**e**); (**h**,**i**) and (**j**,**k**) are, respectively, the GPA results of area-1 and area-3 in (**b**).

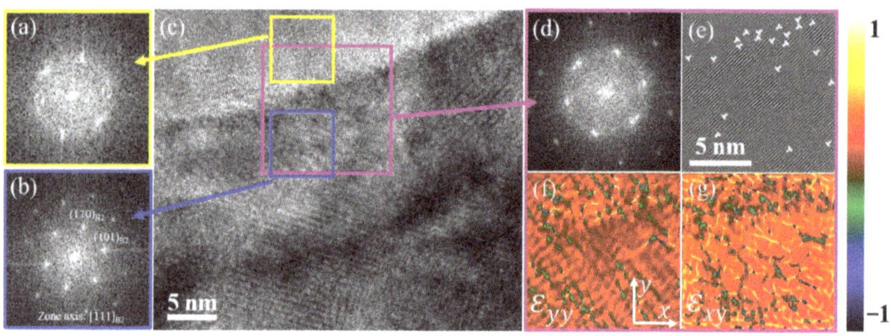

Figure 11. TEM observation of a grain boundary for the trained sample: (**a**,**b**,**d**) show the FFT patterns of the area indicated by yellow, blue and pink squares, respectively; (**c**) the HRTEM image; (**e–g**) show the IFFT patterns and GPA results of the area indicated by pink square.

Figure 11 shows the TEM analysis of a grain boundary. Figure 11a–c shows the FFT patterns for the yellow, blue, and pink square domain in Figure 11c. The FFT pattern of the area indicated by a pink square (Figure 11d) shows a superimposed image of Figure 11a,b, with a small deflection, indicating that this grain boundary was an LAGB. The corresponding IFFT of Figure 11d verified the presence of many dislocations at the grain boundary (Figure 11e). The GPA results (Figure 11f,g) demonstrated that there was a large amount of atomic strain distributed at the grain boundary, which is similar to that in the dislocation wall region. In particular, high-level shear strain existed in the grain boundary region (Figure 11g). The results of the IFFT and GPA of the grain boundary illustrated the existence of dislocations at the grain boundaries, which corresponded to the high KAM value at the interfaces of different grains detected by TKD. The presence of lattice strain indicated that internal stress remained inside the sample after training, leading to the decrease in σ_{SIM} during fatigue tests of the trained sample. On the other hand, the presence of many dislocations inside the grains and at the grain boundaries resulted in the accumulation of residual strain.

Compared to those of the as-received sample, the functional stability and structural fatigue life of the trained sample were effectively improved. The decreased misorientation of the grain boundaries and the FFT analysis both verified that lattice rotation occurred during training, leading to the strong <111> || DD texture. To our knowledge, the excellent fatigue resistance performance resulting from the preferred grain orientation after training has not been previously reported. The relationship between the lattice rotation and enhanced fatigue resistance will be discussed in the next section.

4. Discussion

4.1. *Influence of Training on Functional Degradation*

Generally, functional degradation is caused by dislocation plasticity [32,37], deformation twins [38–40], and residual martensite [32,41,42]. In this study, XRD results (see Figure S2 in Section 4, Supplementary Materials) and TEM observation (Figure 10a–e) indicated that no residual martensite or deformation twins were observed for the trained sample, which is in line with the previous research [9,43]. Therefore, the generation and accumulation of dislocations resulted in functional degradation, including the accumulation of residual strain and a decrease in σ_{SIM} during training.

Plastic deformation involving dislocation activities is generally considered to be a contributor to the evolution of texture [44–46]. During dislocation slipping, the shear stress on the slip plane is divided into two parts, a component stress parallel to the slip direction and component stress vertical to the slip direction. The former is the effective stress that causes dislocations to slip, and the latter tends to cause the crystal planes to rotate. In this study, TEM observations demonstrate that the lattice rotation resulting from training-induced dislocations is responsible for the enhanced <111> || DD texture of the trained sample (Figures 7b,d and 10a–f).

The training-induced preferential grain orientation contributed to reduce the misorientation at the grain boundaries (Figure 8a,b). The incompatible strain at the grain boundaries reduced with the reduction in the misorientation of the adjacent grains, leading to a decrease in plastic deformation in the subsequent fatigue cycles. Besides, according to GPA results, training-induced dislocations caused a local internal stress field inside the sample (shown in Figures 10h–k and 11f,g), which promoted martensite nucleation and lowered the critical external stress needed for martensite transformation during fatigue cycles (Figure 4b). The preferential grain orientation and decreased transformation stress caused by training-induced dislocations reduced the formation and slipping of dislocations, resulting in a lower accumulated residual strain for the trained sample at the same cycles (Figure 4a), i.e., an enhanced cyclic stability.

In this study, the training-induced <111> || DD texture was also a key factor for the improvement of the functional stability. The grains with the $[111]_{B2}$ orientations have a larger transformation strain in the loading direction [47]. Therefore, a smaller volume fraction of B2 grains in the NiTi wires with a <111> || DD texture underwent martensitic transformation upon strain-controlled fatigue cycles. As the volume fraction of the martensite-austenite interface decreased, so did the plastic deformation. In addition, the calculated Taylor factor, which is equal to the inverse of the Schmid factor, has the minimum value for the <111> oriented grains [48]. This result indicated that the stress required for the martensitic transformation of the trained sample with a <111> || DD texture was the lowest. Thus, the driving force for the formation and movement was likewise the lowest, delaying functional degradation. As mentioned above, the suppressed dislocation activity during fatigue tests was the main factor that caused the outstanding functional stability of the trained sample.

4.2. Influence of Training on Structural Fatigue

The enhanced structural fatigue life due to training-induced preferential grain orientation was attributed to the reduced stress concentration and increased energy dissipation ahead of the crack tip. In addition to the dislocations introduced during processing, local stress concentration mainly resulted from incompatible strains near grain boundaries. To determine the strain distribution and induced strain energy during loading, finite element simulations were performed for the as-received and trained samples. The simulations are performed using a finite element multiphysics object-oriented simulation environment (MOOSE) using the model presented in reference [49–51]. The elastic constants and the mechanical properties used in the simulation are shown in Table S5 of supplementary according to ref. [52]. The grain orientations used in the simulations were assigned by the Euler angles extracted from the TKD results (Figure 12a,b). The orientations were indicated by the IPF colors. The tensile loading applied to the top and bottom boundaries was fixed in the y direction. Figure 12c,d presents the distribution of the shear strain (ε_{xy}) and strain in the loading direction (ε_{yy}) of the as-received sample. Section 3 in the Supplementary material shows the calculation details of finite element simulations. For the as-received sample, the local strain concentration at the grain boundary regions was caused by the uncoordinated transformation strain of grains with different orientations; this is in line with the results of the previous report [53]. The presence of shear strain in the as-received sample indicated that the lattice might rotate when under deformation. As for the trained sample, the strong <111> | | DD texture enabled polycrystalline NiTi to exhibit phase transformation behavior similar to that of a single crystal. The grain orientations were highly uniform, so the shear strain was negligible (Figure 12f) and the strain in the loading direction was evenly distributed inside the sample (Figure 12g). The strain concentration near the grain boundaries was significantly reduced. Figure 12e,h shows the strain energy density calculated based on the stress and strain distribution in the as-received and the trained samples, respectively. The strain energy induced by deformation provides the energy needed to create crack surfaces and is thus the driving force of crack propagation [54]. Higher strain energy density in the as-received sample suggested faster crack growth and an earlier failure of the material.

The strong <111> | | DD texture led to an increase in energy dissipation ahead of the crack tip. Sun et al. [18] proposed that energy dissipation due to phase transformation and plastic deformation in front of the crack tip has a shielding effect on the propagation of cracks, and NiTi alloys with larger grains exhibit better shielding against the crack opening, leading to a higher fatigue life. Similar to the large-grain samples, the trained sample with the strong <111> | | DD texture was more prone to plastic deformation [55] and had the lowest σ_{SIM} [48]. Therefore, the sizes of the crack-tip phase transformation zone and the plastic zone were enlarged for the trained sample compared to those of the as-received sample, so the crack-tip shielding effect was enhanced, which increased the resistance to crack growth. Furthermore, the decreased σ_{SIM} value reduced the driving force of crack propagation. The training-induced <111> | | DD texture contributed to reduce the driving force and increase the resistance to crack growth, slowing the crack propagation process and improving the structural fatigue life of NiTi SMA wires.

The fracture surfaces of the as-received and trained samples after fatigue tests until failure are shown in Figure 13. The fracture surfaces for both samples were divided into crack nucleation points, fatigue crack propagation areas, and ductile fracture areas. Surface scratches and impurities introduced during material processing generally act as the crack nucleation sites [56]. In this study, the surface state of both groups of NiTi wires was similar. The growth of cracks for the as-received sample was multi-regional, whereas that of the trained sample was single-regional. The difference was attributed to the effective driving force for crack growth. For the as-received sample, the applied stress was higher during the fatigue test (Figure 4b), and the local stress concentration caused by the anisotropy of the grains with different orientations was more severe; both contributed to the formation of persistent slip bands and micro-cracks. The persistent slip bands reduced the local stress intensity factor, and the micro-cracks evolved into macroscopic cracks [57,58]. Thus, the fatigue crack propagation is easier in the as-received sample. In the trained sample, the relatively uniform stress distribution and increased energy dissipation reduced the driving force and increased the resistance to crack propagation, so the cracks tended to propagate in the preferred direction.

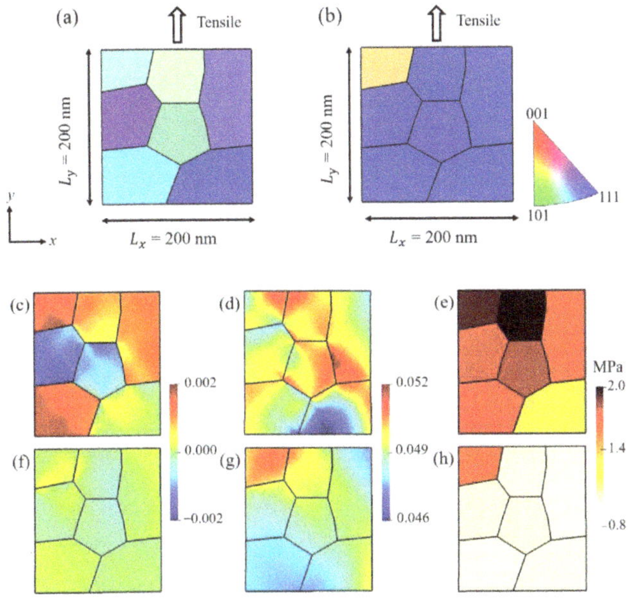

Figure 12. Grain orientations used in the simulations for (**a**) the as-received and (**b**) the trained samples, respectively. (**c**–**e**) are the contour plots of $\varepsilon_{xy}, \varepsilon_{yy}$ and the strain energy density of the as-received sample, respectively. (**f**–**h**) are the contour plots of $\varepsilon_{xy}, \varepsilon_{yy}$ and the strain energy density of the trained sample, respectively. The applied strain was 5%.

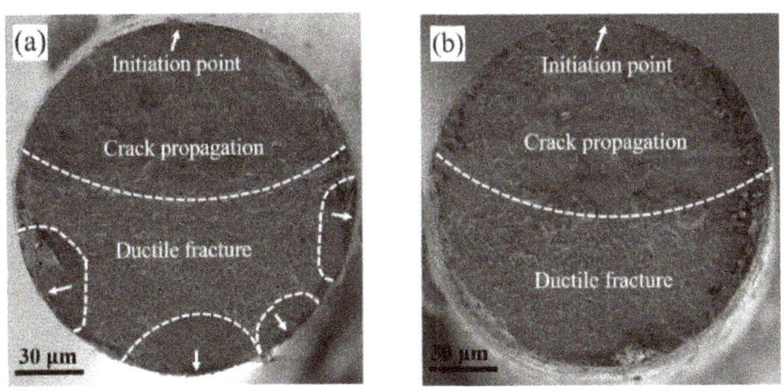

Figure 13. Fracture surface of the nanocrystalline (NC) NiTi wire: (**a**) as-received, and (**b**) trained. The initiation points of the cracks are marked with white arrows.

Crack propagation also has an influence on the functional degradation. As mentioned in Section 3.2, the functional degradation involves two stages during the fatigue tests (Figure 4), which was different from previous reports [6,59–61]. Here, we propose a transition in the fatigue mode during the tests. During the first stage of fatigue (around 1–5000 cycles), functional fatigue dominated the degradation of material performance. Because the stress applied during the fatigue test for the as-received sample was much higher than that of the trained sample in the first stage (Figure 4b), the functional degradation of the as-received sample was faster. At approximately the 5000th cycle, cracks nucleate at the stress concentration regions on the wire surfaces, followed by the propagation of cracks until failure occurred. Thus, structural fatigue dominated the evolution of the functional parameters in the second stage (5000th cycle to failure). The duration of the second stage indicated the difference in propagation processes between the as-received and trained samples. As mentioned above, the effective driving force for crack growth for the trained sample with the <111> | | DD texture was much lower. Thus, the second stage of the trained sample had more cycles. The change in fatigue mode resulted in a two-stage evolution of σ_{SIM}, residual strain, and pseudoelastic strain, and the slower crack propagation caused by the training-induced <111> | | DD texture contributed to a longer structural life.

5. Conclusions

In this study, significantly enhanced fatigue resistance was achieved via performing a stress-controlled training on NC NiTi SMA wires. The associated microscopic mechanism was investigated based on detailed TKD and HRTEM analyses. The conclusions of this work can be summarized as follows:

1. Using a stress-control mechanical training procedure with a peak stress of 600 MPa for 60 cycles at room temperature, the average structural fatigue life increased from 4538 to 13,040 cycles in the strain-control fatigue tests, with the maximum strain of 5%, and the functional stability was also greatly enhanced.
2. Microstructure characterization shows that the <111> texture has been significantly enhanced after training by using the TKD technique. TEM observations demonstrate that the lattice rotation resulting from training-induced dislocations is responsible for the evolution of texture.
3. The improvement in fatigue resistance was attributed to the training-induced preferential grain orientation, which led to the reduction of the driving force and an increase in the resistance to crack propagation. Finite element method simulation results reveal that training-enhanced <111> texture contributes to reduce the incompatible deformation between different grains.

This work proposes a concise and effective mechanical training strategy to enhance the fatigue resistance of NiTi wires, providing new inspiration for the microstructure design of high-performance shape memory alloys.

Supplementary Materials: The following supporting information can be downloaded at: https://www.mdpi.com/article/10.3390/met13020361/s1, Figure S1: The loading-unloading stress-strain curve of uniaxial tension with a peak stress of 700 MPa; Figure S2: X-ray diffraction profiles of the NiTi shape memory alloy wires; Table S1: The grain orientations extracted from the as-received sample; Table S2: The grain orientations extracted from the trained sample; Table S3: The grain orientations extracted from the as-received sample. The Euler angles are in degrees; Table S4: The grain orientations extracted from the trained sample; Table S5: The elastic constants and the mechanical properties used in the simulation.

Author Contributions: Conceptualization, X.C. and M.J.; methodology, P.C. and Y.L.; investigation, P.C., N.M., and Z.W.; writing—original draft preparation, P.C.; writing—review and editing, X.C. and M.J.; project administration, X.J. All authors have read and agreed to the published version of the manuscript.

Funding: This work was supported by the National Natural Science Foundation of China (grant number 51201100).

Institutional Review Board Statement: Not applicable.

Data Availability Statement: Data are available from the corresponding author on reasonable request.

Acknowledgments: We thank Yongjia Liu from the Instrumental Analysis Center of Shanghai Jiao Tong University for assistance with the mechanical performance testing of DMA.

Conflicts of Interest: The authors declare no conflict of interest.

References

1. Shabalovskaya, S.A. On the nature of the biocompatibility and on medical applications of NiTi shape memory and superelastic alloys. *Bio-Med. Mater. Eng.* **1996**, *6*, 267–289. [CrossRef]
2. Drukteinis, S.; Peciuliene, V.; Bendinskaite, R.; Brukiene, V.; Maneliene, R.; Nedzinskiene, E. Shaping Ability, Cyclic Fatigue Resistance and Fractographic Analysis of Hybrid and Reciprocating Nickel Titanium Endodontic Instruments. *Metals* **2020**, *10*, 172. [CrossRef]
3. Hartl, D.J.; Lagoudas, D. Aerospace applications of shape memory alloys. *Proc. Inst. Mech. Eng. Part G J. Aerosp. Eng.* **2014**, *221*, 535–552. [CrossRef]
4. Mohd-Jani, J.; Leary, M.; Subic, A.; Gibson, M.A. A review of shape memory alloy research, applications and opportunities. *Mater. Des.* **2014**, *56*, 1078–1113. [CrossRef]
5. Zhang, Y.; You, Y.; Moumni, Z.; Anlas, G.; Zhu, J.; Zhang, W. Experimental and theoretical investigation of the frequency effect on low cycle fatigue of shape memory alloys. *Int. J. Plast.* **2017**, *90*, 1–30. [CrossRef]
6. Zhang, Y.; Moumni, Z.; Zhu, J.; Zhang, W. Effect of the amplitude of the training stress on the fatigue lifetime of NiTi shape memory alloys. *Scr. Mater.* **2018**, *149*, 66–69. [CrossRef]
7. Eggeler, G.; Hornbogen, E.; Yawny, A.; Heckmann, A.; Wagner, M. Structural and functional fatigue of NiTi shape memory alloys. *Mater. Sci. Eng. A* **2004**, *378*, 24–33. [CrossRef]
8. Paranjape, H.M.; Bowers, M.L.; Mills, M.J.; Anderson, P.M. Mechanisms for phase transformation induced slip in shape memory alloy micro-crystals. *Acta Mater.* **2017**, *132*, 444–454. [CrossRef]
9. Delville, R.; Malard, B.; Pilch, J.; Sittner, P.; Schryvers, D. Transmission electron microscopy investigation of dislocation slip during superelastic cycling of Ni–Ti wires. *Int. J. Plast.* **2011**, *27*, 282–297. [CrossRef]
10. Delville, R.; Malard, B.; Pilch, J.; Sittner, P.; Schryvers, D. Microstructure changes during non-conventional heat treatment of thin Ni–Ti wires by pulsed electric current studied by transmission electron microscopy. *Acta Mater.* **2010**, *58*, 4503–4515. [CrossRef]
11. Ahadi, A.; Sun, Q. Stress-induced nanoscale phase transition in superelastic NiTi by in situ X-ray diffraction. *Acta Mater.* **2015**, *90*, 272–281. [CrossRef]
12. Robertson, S.W.; Pelton, A.R.; Ritchie, R.O. Mechanical fatigue and fracture of Nitinol. *Int. Mater. Rev.* **2013**, *57*, 1–37. [CrossRef]
13. Glushko, O.; Kiener, D. Initiation of fatigue damage in ultrafine grained metal films. *Acta Mater.* **2021**, *206*, 116599. [CrossRef]
14. Gollerthan, S.; Young, M.; Baruj, A.; Frenzel, J.; Schmahl, W.; Eggeler, G. Fracture mechanics and microstructure in NiTi shape memory alloys. *Acta Mater.* **2009**, *57*, 1015–1025. [CrossRef]
15. Robertson, S.; Mehta, A.; Pelton, A.; Ritchie, R. Evolution of crack-tip transformation zones in superelastic Nitinol subjected to in situ fatigue: A fracture mechanics and synchrotron X-ray microdiffraction analysis. *Acta Mater.* **2007**, *55*, 6198–6207. [CrossRef]
16. Tyc, O.; Molnárová, O.; Šittner, P. Effect of microstructure on fatigue of superelastic NiTi wires. *Int. J. Fatigue* **2021**, *152*, 106400. [CrossRef]
17. Zhang, K.; Kang, G.; Sun, Q. High fatigue life and cooling efficiency of NiTi shape memory alloy under cyclic compression. *Scr. Mater.* **2019**, *159*, 62–67. [CrossRef]
18. Chen, J.; Yin, H.; Sun, Q. Effects of grain size on fatigue crack growth behaviors of nanocrystalline superelastic NiTi shape memory alloys. *Acta Mater.* **2020**, *195*, 141–150. [CrossRef]
19. Xie, G.; Wang, F.; Song, B.; Cheng, J.; Wang, J.; Zeng, X. Grain size dependence of cracking performance in polycrystalline NiTi alloys. *J. Alloy. Compd.* **2021**, *884*, 161132. [CrossRef]
20. Yin, H.; He, Y.; Moumni, Z.; Sun, Q. Effects of grain size on tensile fatigue life of nanostructured NiTi shape memory alloy. *Int. J. Fatigue* **2016**, *88*, 166–177. [CrossRef]
21. Karakoc, O.; Atli, K.; Evirgen, A.; Pons, J.; Santamarta, R.; Benafan, O.; Noebe, R.; Karaman, I. Effects of training on the thermomechanical behavior of NiTiHf and NiTiZr high temperature shape memory alloys. *Mater. Sci. Eng. A* **2020**, *794*, 139857. [CrossRef]
22. Hayrettin, C.; Karakoc, O.; Karaman, I.; Mabe, J.; Santamarta, R.; Pons, J. Two way shape memory effect in NiTiHf high temperature shape memory alloy tubes. *Acta Mater.* **2019**, *163*, 1–13. [CrossRef]
23. Chang, C.-Y.; Vokoun, D.; Hu, C.-T. Two-way shape memory effect of NiTi alloy induced by constraint aging treatment at room temperature. *Met. Mater. Trans. A* **2001**, *32*, 1629–1634. [CrossRef]
24. Chu, K.; Sun, Q. Reducing functional fatigue, transition stress and hysteresis of NiTi micropillars by one-step overstressed plastic deformation. *Scr. Mater.* **2021**, *201*, 113958. [CrossRef]
25. Perkins, J.; Bobowiec, P. Microstructural effects of martensitic transformation cycling of a Cu-Zn-Al alloy: Vestigial structures in the parent phase. *Met. Mater. Trans. A* **1986**, *17*, 195–203. [CrossRef]
26. Trimby, P.W. Orientation mapping of nanostructured materials using transmission Kikuchi diffraction in the scanning electron microscope. *Ultramicroscopy* **2012**, *120*, 16–24. [CrossRef] [PubMed]

27. Trimby, P.W.; Cao, Y.; Chen, Z.; Han, S.; Hemker, K.J.; Lian, J.; Liao, X.; Rottmann, P.; Samudrala, S.; Sun, J.; et al. Characterizing deformed ultrafine-grained and nanocrystalline materials using transmission Kikuchi diffraction in a scanning electron microscope. *Acta Mater.* **2014**, *62*, 69–80. [CrossRef]
28. Jaureguizahar, S.M.; Chapetti, M.D.; Yawny, A.A. Fatigue of NiTi shape memory wires. *Procedia Struct. Integr.* **2016**, *2*, 1427–1434. [CrossRef]
29. Tyc, O.; Heller, L.; Vronka, M.; Šittner, P. Effect of temperature on fatigue of superelastic NiTi wires. *Int. J. Fatigue* **2020**, *134*, 105470. [CrossRef]
30. Tyc, O.; Pilch, J.; Sittner, P. Fatigue of superelastic NiTi wires with different plateau strain. *Procedia Struct. Integr.* **2016**, *2*, 1489–1496. [CrossRef]
31. Zhang, Y.; Moumni, Z.; You, Y.; Zhang, W.; Zhu, J.; Anlas, G. Multiscale TRIP-based investigation of low-cycle fatigue of polycrystalline NiTi shape memory alloys. *Int. J. Plast.* **2018**, *115*, 307–329. [CrossRef]
32. Hua, P.; Lin, H.; Sun, Q. Ultrahigh cycle fatigue deformation of polycrystalline NiTi micropillars. *Scr. Mater.* **2021**, *203*, 114108. [CrossRef]
33. Cai, X.; Handwerker, C.A.; Blendell, J.E.; Koslowski, M. Shallow grain formation in Sn thin films. *Acta Mater.* **2020**, *192*, 1–10. [CrossRef]
34. Jahedi, M.; Ardeljan, M.; Beyerlein, I.J.; Paydar, M.H.; Knezevic, M. Enhancement of orientation gradients during simple shear deformation by application of simple compression. *J. Appl. Phys.* **2015**, *117*, 214309. [CrossRef]
35. Motz, C.; Schöberl, T.; Pippan, R. Mechanical properties of micro-sized copper bending beams machined by the focused ion beam technique. *Acta Mater.* **2005**, *53*, 4269–4279. [CrossRef]
36. Gao, H.; Huang, Y.; Nix, W.; Hutchinson, J. Mechanism-based strain gradient plasticity? I. Theory. *J. Mech. Phys. Solids* **1999**, *47*, 1239–1263. [CrossRef]
37. Xiao, Y.; Zeng, P.; Lei, L.; Zhang, Y. In situ observation on temperature dependence of martensitic transformation and plastic deformation in superelastic NiTi shape memory alloy. *Mater. Des.* **2017**, *134*, 111–120. [CrossRef]
38. Gao, Y.; Casalena, L.; Bowers, M.; Noebe, R.; Mills, M.; Wang, Y. An origin of functional fatigue of shape memory alloys. *Acta Mater.* **2017**, *126*, 389–400. [CrossRef]
39. Šittner, P.; Molnárová, O.; Kaderavek, L.; Tyc, O.; Heller, L. Deformation twinning in martensite affecting functional behavior of NiTi shape memory alloys. *Materialia* **2020**, *9*, 100506. [CrossRef]
40. Chen, Y.; Tyc, O.; Kadeřávek, L.; Molnárová, O.; Heller, L.; Šittner, P. Temperature and microstructure dependence of localized tensile deformation of superelastic NiTi wires. *Mater. Des.* **2019**, *174*, 107797. [CrossRef]
41. Polatidis, E.; Šmíd, M.; Kuběna, I.; Hsu, W.-N.; Laplanche, G.; Van Swygenhoven, H. Deformation mechanisms in a superelastic NiTi alloy: An in-situ high resolution digital image correlation study. *Mater. Des.* **2020**, *191*, 108622. [CrossRef]
42. Belyaev, S.; Resnina, N.; Ponikarova, I.; Iaparova, E.; Rakhimov, T.; Ivanova, A.; Tabachkova, N.; Andreev, V. Damage of the martensite interfaces as the mechanism of the martensite stabilization effect in the NiTi shape memory alloys. *J. Alloy. Compd.* **2022**, *921*, 166189. [CrossRef]
43. Polatidis, E.; Zotov, N.; Bischoff, E.; Mittemeijer, E. The effect of cyclic tensile loading on the stress-induced transformation mechanism in superelastic NiTi alloys: An in-situ X-ray diffraction study. *Scr. Mater.* **2015**, *100*, 59–62. [CrossRef]
44. Li, H.; Chen, T.; Li, W.; Zhang, H.; Han, S.; Zhou, C.; Chen, Z.; Flores-Johnson, E.A.; Shen, L.; Lian, J.; et al. Grain size dependent microstructure and texture evolution during dynamic deformation of nanocrystalline face-centered cubic materials. *Acta Mater.* **2021**, *216*, 117088. [CrossRef]
45. Madhavan, R.; Ray, R.; Suwas, S. Texture transition in cold-rolled nickel–40wt.% cobalt alloy. *Acta Mater.* **2014**, *74*, 151–164. [CrossRef]
46. Madhavan, R.; Suwas, S. Micro-mechanisms of deformation texture evolution in nanocrystalline nickel-cobalt alloys. *Acta Mater.* **2016**, *121*, 46–58. [CrossRef]
47. Buchheit, T.E.; Wert, J.A. Predicting the orientation-dependent stress-induced transformation and detwinning response of shape memory alloy single crystals. *Met. Mater. Trans. A* **1996**, *27*, 269–279. [CrossRef]
48. Ono, N.; Satoh, A.; Ohta, H. A Discussion on the Mechanical Properties of Shape Memory Alloys Based on a Polycrystal Model. *Mater. Trans. JIM* **1989**, *30*, 756–764. [CrossRef]
49. Alger, B.; Andrš, D.; Carlsen, R.W.; Gaston, D.R.; Kong, F.; Lindsay, A.D.; Miller, J.M.; Permann, C.J.; Peterson, J.W.; Slaughter, A.E.; et al. MOOSE Web Page. 2020. Available online: https://mooseframework.org (accessed on 2 August 2021).
50. Simo, J.C.; Hughes, T.J.R. *Computational Inelasticity: Volume 7 of Interdisciplinary Applied Mathematics*; Springer: Secaucus, NJ, USA, 1998.
51. Xu, L.; Solomou, A.; Baxevanis, T.; Lagoudas, D. Finite strain constitutive modeling for shape memory alloys considering transformation-induced plasticity and two-way shape memory effect. *Int. J. Solids Struct.* **2020**, *221*, 42–59. [CrossRef]
52. Wagner, M.F.-X.; Windl, W. Elastic anisotropy of Ni4Ti3 from first principles. *Scr. Mater.* **2009**, *60*, 207–210. [CrossRef]
53. Lackmann, J.; Niendorf, T.; Maxisch, M.; Grundmeier, G.; Maier, H. High-resolution in-situ characterization of the surface evolution of a polycrystalline NiTi SMA-alloy under pseudoelastic deformation. *Mater. Charact.* **2011**, *62*, 298–303. [CrossRef]
54. Griffith, A.A. VI. The phenomena of rupture and flow in solids. *Philos. Trans. R. Soc. Lond. Ser. A* **1920**, *221*, 163–198.
55. Paranjape, H.; Anderson, P.M. Texture and grain neighborhood effects on Ni–Ti shape memory alloy performance. *Model. Simul. Mater. Sci. Eng.* **2014**, *22*, 075002. [CrossRef]

56. Rahim, M.; Frenzel, J.; Frotscher, M.; Pfetzing-Micklich, J.; Steegmüller, R.; Wohlschlögel, M.; Mughrabi, H.; Eggeler, G. Impurity levels and fatigue lives of pseudoelastic NiTi shape memory alloys. *Acta Mater.* **2013**, *61*, 3667–3686. [CrossRef]
57. Cervellon, A.; Hémery, S.; Kürnsteiner, P.; Gault, B.; Kontis, P.; Cormier, J. Crack initiation mechanisms during very high cycle fatigue of Ni-based single crystal superalloys at high temperature. *Acta Mater.* **2020**, *188*, 131–144. [CrossRef]
58. Frenzel, J. On the Importance of Structural and Functional Fatigue in Shape Memory Technology. *Shape Mem. Superelasticity* **2020**, *6*, 213–222. [CrossRef]
59. Hua, P.; Chu, K.; Ren, F.; Sun, Q. Cyclic phase transformation behavior of nanocrystalline NiTi at microscale. *Acta Mater.* **2020**, *185*, 507–517. [CrossRef]
60. Karakoc, O.; Hayrettin, C.; Canadinc, D.; Karaman, I. Role of applied stress level on the actuation fatigue behavior of NiTiHf high temperature shape memory alloys. *Acta Mater.* **2018**, *153*, 156–168. [CrossRef]
61. Hou, H.; Simsek, E.; Ma, T.; Johnson, N.S.; Qian, S.; Cissé, C.; Stasak, D.; Al Hasan, N.; Zhou, L.; Hwang, Y.; et al. Fatigue-resistant high-performance elastocaloric materials made by additive manufacturing. *Science* **2019**, *366*, 1116–1121. [CrossRef]

metals

MDPI

Article

Large Magnetic Entropy Change in GdRuSi Optimal for Magnetocaloric Liquefaction of Nitrogen

Anatoly G. Kuchin [1], Sergey P. Platonov [1], Roman D. Mukhachev [1], Alexey V. Lukoyanov [1,2,*], Aleksey S. Volegov [1,3], Vasilii S. Gaviko [1,3] and Mari Yu. Yakovleva [1]

1 M.N. Mikheev Institute of Metal Physics of Ural Branch of Russian Academy of Sciences, 620108 Ekaterinburg, Russia
2 Institute of Physics and Technology, Ural Federal University Named after the First President of Russia B.N. Yeltsin, 620002 Ekaterinburg, Russia
3 Institute of Natural Sciences and Mathematics, Ural Federal University Named after the First President of Russia B.N. Yeltsin, 620026 Ekaterinburg, Russia
* Correspondence: lukoyanov@imp.uran.ru; Tel.: +7-3433783886

Abstract: The search for new magnetocaloric materials for application in magnetic refrigerants is highly motivated by high efficiency, reliability, and environmental safety. The values of the magnetocaloric effect MCE and the refrigerant capacity *RC* for the equiatomic GdRuSi compound were determined to be MCE = 10.7 and 4.94 J/kgK and *RC* = 336 and 92 J/kg with a change in the field of 0–50 and 0–17 kOe, respectively. These high values of MCE and *RC* for GdRuSi appear in the region of nitrogen liquefaction temperature of 77.4 K, due to which the compound can be useful in practice. The densities of states and magnetic moments of GdRuSi were calculated theoretically, taking into account strong electron correlations in the 4f Gd shell. The total magnetic moment of GdRuSi was found to be composed of the Gd moment only with the value calculated in very good agreement with the experimental one.

Keywords: magnetic measurements; magnetocaloric effect; electronic structure; intermetallics

1. Introduction

Functional magnetic materials such as permanent or soft magnets, magnetostrictive, and magnetic shape memory compounds are very important for application in modern society. An urgent problem is the search for new magnetocaloric materials for magnetic refrigerates in view of high efficiency, reliability, and environmental safety [1–3]. The outstanding magnetocaloric performances have been reported on heavy rare earth-based compounds [1–5], MnTZ (T = Co and Fe; Z = Ge and Si)-based compounds [6], Ni-Mn-Y (Y = Ga, In and Sn)-based compounds [7], La(Fe, $Si)_{13}H_x$-based compounds [8], etc., with the MCE about 20–25 J/kgK in a field changing to 50 kOe. The temperature range in which MCE is pronounced is also important [1–4]. The so-called refrigerant capacity *RC* indicates how much heat can be transferred from the cold end to the hot end of a refrigerator in one refrigeration cycle. The giant-MCE materials have large MCE values extending over a narrow temperature range and irreversible character of MCE with a hysteresis due to the first-order phase transitions. Materials with the second-order phase transitions (the most used is Gd with MCE = 9.8 J/kgK at T_C = 293 K in a field changing to 50 kOe) present smaller MCE values spread over substantially a broader temperature range. It should be noted that the vast majority of the magnetocaloric materials exhibit large MCE values of the order of 10–30 J/kgK at cryogenic temperatures [9,10].

The ternary intermetallic RTX compounds (R = rare earth, T = transitional metal, X = p block metal) have attracted much attention, due to their interesting physical properties [9]. The RTX compounds have gained significant importance in power, electronics, and telecommunications industries because of their magnetic, electronic, and chemical properties [9]. Recently, a ternary Gd-Ru-Si intermetallic compound with the close 122 composition attracted tremendous attention from researchers due to the nanometric square skyrmion lattice discovered in its centrosymmetric tetragonal lattice [11].

The GdFeSi and GdRuSi compounds crystallize in tetragonal structure of the CeFeSi type (space group *P4/nmm*) [12,13]. The hybridization between Si p states and 3d states of transition metal atom leads to the filling of 3d band and absence of the magnetic moment of Fe in the RFeSi compounds [12]. The GdFeSi and GdRuSi compounds are ferromagnetic below T_C = 135 and 85 K, respectively [12,13]. It was found that the GdFeSi compound exhibits a rather large value of the magnetocaloric effect MCE of 5 J/kgK [14,15] at T_C = 130 K [15] in a field changing to 17 kOe. This value is in line with the following new promising rare-earth-based compounds with a hysteresis-free second-order magnetic phase transition at about 100 K for use in magnetic refrigeration, such as DyFeSi with maximum magnetic entropy change (i.e., MCE) $-\Delta S_M$ = 9.2 J/kgK at 70 K with the magnetic field change of 0–20 kOe [16], TbFeSi with $-\Delta S_M$ = 9.8 J/kgK (0–20 kOe) at 110 K [16], Tb_3Co with $-\Delta S_M$ = 28.8 J/kgK (0–70 kOe) at 84 K [17], Gd_2Cu_2Cd with $-\Delta S_M$ = 7.8 J/kgK (0–50 kOe) at 120 K [18]. We can assume the same large value of the MCE in the case of GdRuSi as for GdFeSi, keeping in mind the close values of the MCE for ErRuSi (21.2 J/kgK) and ErFeSi (23.1 J/kgK) in a field changing to 50 kOe [9] and that only the R component is magnetic in RRuSi and RFeSi [9,12,13]. Considering the proximity of T_C = 85 K of GdRuSi [9] to the liquefaction of nitrogen temperature T = 77.4 K, it is of interest to study the magnetocaloric effect for GdRuSi.

2. Materials and Methods

The GdRuSi compound was prepared by arc melting in an argon atmosphere. The ingot was homogenized at 1273 K for 8 days and then quenched in water. X-ray powder diffraction analysis was employed to determine the phase composition, structure type, and lattice parameters under ambient conditions. A diffractometer of Empyrean Series 2 (PANalytical, Malvern, UK) and CuKα radiation was used. The HighScore v.4.x programs were employed for the calculation of lattice parameters and analysis of phase compositions. A vibromagnetometer 7407 VSM (Lake Shore Cryotronics, Westerville, OH, USA) and SQUID-magnetometer MPMS XL7 (Quantum Design, San Diego, CA, USA) were used for magnetic study in a magnetic field up to 7 T. The magnetization curves *M(H)* were measured at 4 K on free powder samples previously oriented by an external field of 7 T upon cooling from room temperature. They were corrected by simple subtraction of the $GdRu_2Si_2$ impurity contribution. The temperature of the magnetic phase transition in the sample was determined from the magnetization vs. temperature $M(T)$ curves measured in a magnetic field of 0.01 T and in the temperature range 4–440 K. The T_C was also determined with the help of Arrott plots.

The first principles theoretical calculations of magnetic moments and electronic structure of the ternary GdRuSi compound were completed within the DFT+U method [19] in the Quantum ESPRESSO computational package [20,21] based on the exchange–correlation functional approximation in the form of generalized gradient approximation (GGA), version Perdew–Burke–Ernzerhof (PBE) [22]. For the sufficient convergence in our self-consistency cycle, an energy cutoff for plane waves equal to 60 Ry was set. Integration in reciprocal space was based on a grid of 12 $\times$ 12 $\times$ 12 **k**-points. In the calculations, we used the projected augmented wave (PAW) datasets for Gd [23] and standard PAW pseudopotentials for Ru and Si from the pseudopotential library of Quantum ESPRESSO [24]. In 4f Gd shell, strong electron correlations were taken into account as the +U correction for direct Coulomb parameter equal to 6.7 eV and exchange parameter equal to 0.7 eV, which are commonly employed for Gd as a metal and in various Gd-based intermetallic compounds [19,25].

3. Results

3.1. Crystal Structure Analysis

The GdRuSi compound investigated crystallize into the tetragonal CeFeSi (*P4/nmm*)–type structure with the lattice parameters a = 4.192 Å and c = 6.602 Å. These values are close to a = 4.173(2) Å and c = 6.586(4) Å found in [13]. The fraction of a parasitic antiferromagnetic $GdRu_2Si_2$ phase is also presented in the alloy, however the amount of dominant phase could be estimated as >95%. The experimental and calculated X-ray diffraction patterns for GdRuSi are shown in Figure 1. The presence of small additional reflections from $GdRu_2Si_2$ were taken into account in simulations.

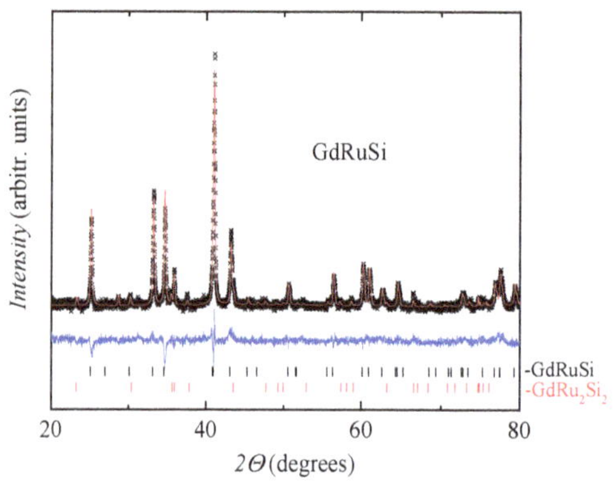

Figure 1. X-ray diffraction pattern (symbols) and its fitting (envelop line) of GdRuSi. The lower curve is the difference between the experimental and calculated results. Vertical lines show the reflection positions for the (up-down) GdRuSi and $GdRu_2Si_2$ compounds.

3.2. Magnetic Properties and Electronic Structure

Figure 2 shows the magnetization curves for the GdRuSi compound, which were measured at 4 K on free powder samples previously oriented along easy magnetization axis by an external field of 70 kOe upon cooling from room temperature. The magnetization curve is typical of ferro- or ferrimagnets, similarly to GdFeSi [25]. The GdRuSi compound has a saturation magnetization of the order of 7 μ_B/f.u. (Figure 2), which equals to the value of the magnetic moment for a free Gd ion. This means it is a collinear ferromagnet. There is virtually no magnetic hysteresis on the M(H) dependence for GdRuSi.

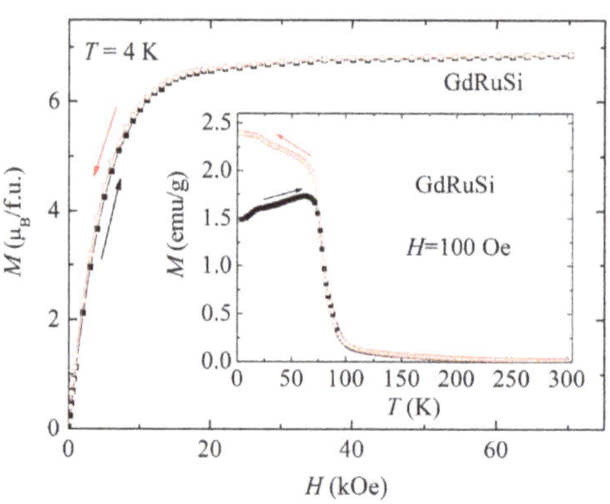

Figure 2. Magnetization and demagnetization curves of free powder sample of GdRuSi at T = 4 K. The inset shows field-cooled (FC) and zero-field-cooled (ZFC) thermomagnetic curves of GdRuSi collected in an applied field of 100 Oe applied along easy magnetization axis of the free powder sample.

The zero-field-cooled thermomagnetic curve ZFC *M(T)* for the GdRuSi compound measured in a field of 100 Oe on free powder sample previously oriented along easy magnetization axis by an external field of 70 kOe is presented in Figure 2. The increase in ZFC *M(T)* upon heating from 4 K can be caused by domain-wall movement [26]. Such an increase in field-cooled FC *M(T)* was not observed when cooling the sample in a field of 100 Oe (see Figure 2) because the maximum order degree for that magnetic field has already been reached [26]. There is no thermal hysteresis on the *M(T)* dependence for GdRuSi in the region of the temperature of the magnetic phase transition of the ferro-paramagnet, i.e., this is a second-order phase transition. The Curie temperature T_C = 78.3 K of the alloys was determined from the position of the minimum of the derivative for the *M(T)* dependence. Earlier in [13] a slightly larger value of T_C = 85 K was established, perhaps because of measurements in higher field up to 15 kOe, which creates an additional ordering effect on the magnetic moments of the ions. It is also seen (Figure 2) that there is a small jump at T~18 K, presumably corresponding to the magnetic phase transition from antiferromagnetic to ferromagnetic ordering in the parasitic $GdRu_2Si_2$ phase [27]. The presence of this phase in the GdRuSi sample was established by X-ray diffraction (see Section 3.1). There are no other features on the *M(T)* dependence caused by magnetic phase transformations in the GdRuSi sample.

The T_C = 78.3 K for GdRuSi was also determined with the help of Arrott plots [28] as is shown in Figure 3. The positive slope of the Arrott curves indicate that the ferromagnet-to-paramagnet transition can be classified as second-order type-phase transition according to the Banerjee criterion [28]. A similar conclusion follows from the absence of thermal hysteresis on the *M(T)* dependence for GdRuSi in the region of the temperature of the ferromagnet-to-paramagnet phase transition (Figure 2) and from the "caret" shape of the temperature dependence of the change in magnetic entropy $-\Delta S_M(T)$ (see Section 3.1), which is typical of the second-order magnetic-phase transition [28].

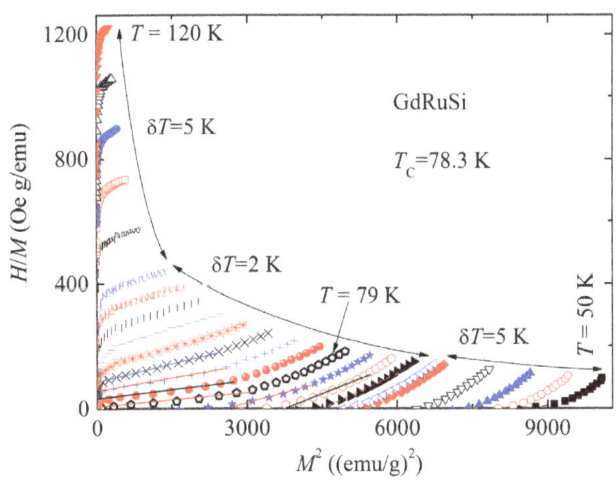

Figure 3. Arrot plots for the GdRuSi compound. Different colors for different symbols are made to better distinguish them.

The densities of electronic states for the GdRuSi compound are shown in Figure 4. The Fermi energy (E_F) is shown as a vertical dashed line at zero energy. The most intense DOS peaks (shown in red in Figure 4) near −6.5 and 5.8 eV are formed by the 4f localized electronic states of Gd. Due to the strong electron correlations in the 4f Gd shell taking into account in our GGA+U calculations, these peaks are shifted from the Fermi energy, because the 4f electronic states are half-filled. In the energy range from −4 eV up to 4 eV, the density of electronic states formed due to the 4d states of Ru and 3p states of Si. These electronic states are almost spin unpolarized, i.e., almost the same for both spin projections. In the calculations, the ferromagnetic ordering of the Gd magnetic moments is found, whereas Ru and Si are converged to be nonmagnetic (less than 0.02 μ_B). Nonmagnetic 3d ions are also found in the calculations for similar Gd-based ternary intermetallics [25,29,30]. For example, in GdTiSi [29] or in the close compound GdNiGe [30], the Ti and Ni, as well as Si or Ge, are calculated to be nonmagnetic, correspondingly. Moreover, in GdFeSi, the electronic structure calculations also revealed that the iron ions are nonmagnetic, which is caused by the presence of silicon in GdFeSi [25]. For the Gd ions, the strong spin polarization at half-filling of the 4f shell results in the magnetic moment 7.1 μ_B (per Gd ion and per f.u.), which is in very good agreement with the above experimental value 7 μ_B/f.u., see above.

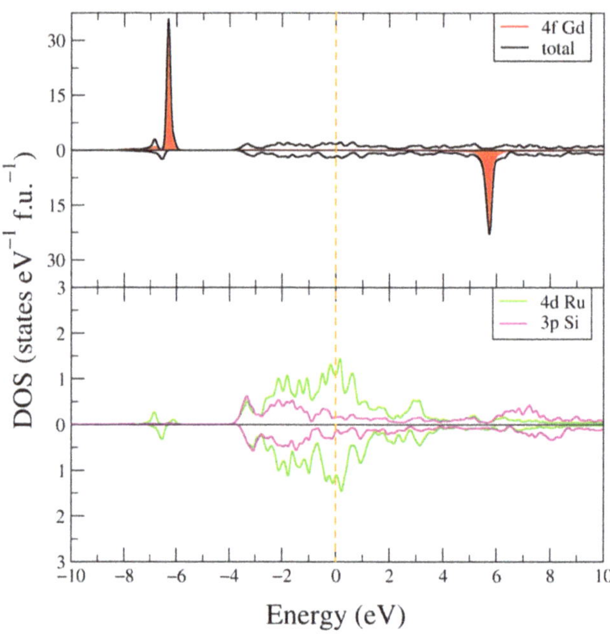

Figure 4. Calculated total (black curves) and partial densities of states of GdRuSi. The Fermi energy (E_F) is set at zero and shown as a vertical dashed orange line.

3.3. Magnetocaloric Effect

The isothermal magnetic entropy change $-\Delta S_M$ (i.e., the magnetocaloric effect MCE) was calculated from the magnetization isotherms *M*(*H*) using the well-known Maxwell relation [9]:

$$\Delta S_M(T,H) = \int_0^H \left(\frac{\partial M}{\partial T}\right)_H dH \quad (1)$$

Magnetization isotherms *M(H)* of GdRuSi in the range *T* = 50–120 K are shown in Figure 5. The temperature step is 5 K in the ranges 50–70, 95–120 K and 2 K in the range 71–95 K.

Figure 6 illustrates the isothermal magnetic entropy changes $-\Delta S_M$ (1) for GdRuSi as a function of temperature in a field changing to 10–50 kOe. Their "caret" shape is typical of the second-order magnetic-phase transition [28]. A similar conclusion follows from the form of the Arrott plots for GdRuSi (Figure 3) and from the absence of thermal hysteresis on the *M(T)* dependence for GdRuSi in the region of the temperature of the ferromagnet-to-paramagnet phase transition (Figure 2).

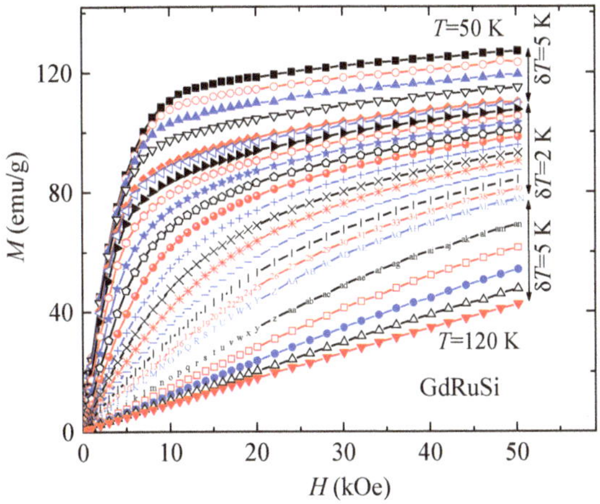

Figure 5. Magnetization isotherms *M(H)* of GdRuSi in the range T = 50–120 K (up-down). The temperature step is 5 K in the ranges 50–70, 95–120 K and 2 K in the range 71–95 K. Different colors for different symbols are made to better distinguish them.

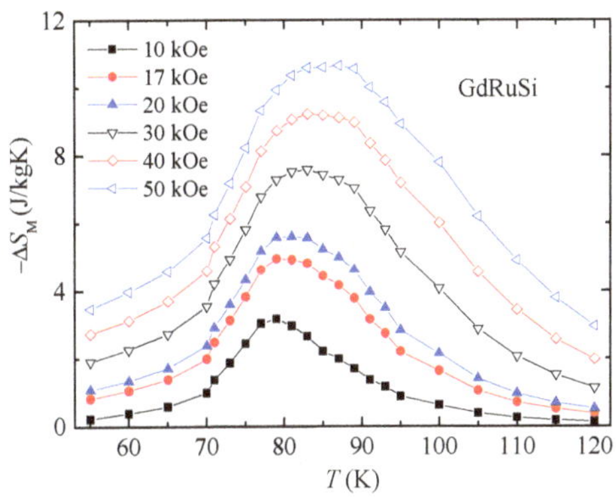

Figure 6. Temperature dependences of the change in magnetic entropy $-\Delta S_M(T)$ in a field changing to 10–50 kOe for GdRuSi.

For the GdRuSi equiatomic compound, a large value of the MCE was found, which is 10.7 at 87 K and 4.94 J/kgK at 79 K in a field change of 0–50 and 0–17 kOe, respectively. The value 4.94 J/kgK is somewhat less than 5 J/kgK for GdFeSi when the field changes 0–17 kOe [15]. At the same time, the Curie temperature T_C of GdFeSi equals to 130 K [15] and of GdRuSi—78.3 K, well below room temperature for practical application in domestic magnetic refrigerator. In this situation, successful alloying can be a good way to increase T_C of these compounds. For example, a sharp increase in T_C up to 184.4 K (by 54.4 K) was established when Ti atoms were substituted for Fe in the $GdFe_{1-x}Ti_xSi$ compounds at $x = 0.1$ [15].

The value $-\Delta S_M(T_C)$ = 5.6 J/kgK for GdRuSi (Figure 6) is noticeably smaller than $-\Delta S_M(T_C)$ = 9.2 J/kgK for DyFeSi [10] in a field changing to 20 kOe. However, T_C = 78.3 K and, consequently, the $-\Delta S_M(T_C)$ maximum for GdRuSi are closer to the nitrogen liquefaction temperature of 77.4 K than T_C = 70 K for DyFeSi. Therefore, GdRuSi is more efficient for nitrogen liquefaction than DyFeSi.

The refrigerant capacity *RC* was estimated by numerical integration of the area under the $-\Delta S_M(T)$ curve between the temperatures T_1 and T_2 at the width of half maximum of the peak [14]:

$$RC = -\int_{T_1}^{T_2} \Delta S_M dT \quad (2)$$

The *RC* value of the GdRuSi compound equals to 336 J/kg (for T_1 = 69 K and T_2 = 109 K) and 92 J/kg (for T_1 = 71 K and T_2 = 94 K) in a field changing to 50 or 17 kOe, respectively.

The GdRuSi compound with large $-\Delta S_M(T_C)$ = 4.94 J/kgK at T_C = 78.3 K and *RC* = 92 J/kg in a field changing to 1.7 T can be of practical interest due to T_C being close to nitrogen liquefaction temperature of 77.4 K.

4. Conclusions

The values of the magnetocaloric effect MCE and the refrigerant capacity *RC* for the equiatomic GdRuSi alloy were determined, which are MCE = 10.7 and 4.94 J/kgK and *RC* = 336 and 92 J/kg with a change in the field of 0–50 and 0–17 kOe, respectively. These large values of MCE and *RC* for GdRuSi with T_C = 78.3 K appear in the region of nitrogen liquefaction temperature of 77.4 K, due to which the compound can be useful in practice. The value of 4.94 J/kgK is quite large and almost coincides with the previously obtained 5 J/kgK at T_C = 130 K and a field change of 0–1.7 T for GdFeSi. In the calculated densities of states of GdRuSi, we identified the localized 4f Gd states, which required the accounting for electron correlations within DFT+U. The theoretical total magnetic moment of GdRuSi was found to be solely formed by the Gd ion (the Ru and Si ions were calculated to be nonmagnetic similar to Fe and Si in the previously considered ternary GdFeSi intermetallic compound), and its value is in very good agreement with the experimental data. Further search of magnetocaloric materials among gadolinium compounds may result in the discovery of novel materials useful for various environmentally sustainable applications.

Author Contributions: Conceptualization; methodology, A.G.K. and A.V.L.; software, R.D.M. and A.V.L.; validation, A.G.K. and A.V.L.; investigation, S.P.P., R.D.M., A.V.L., A.S.V. and V.S.G.; formal analysis, V.S.G. and S.P.P.; data curation; visualization, M.Y.Y.; writing—original draft preparation; writing—review and editing, A.G.K. and A.V.L.; supervision, A.G.K. and A.V.L.; project administration, A.V.L. All authors have read and agreed to the published version of the manuscript.

Funding: The results of Sections 1, 2, 3.1 and 3.2 (magnetic properties and electronic structure calculations) were obtained within the grant of the Russian Science Foundation No. 18-72-10098, https://rscf.ru/en/project/18-72-10098/ (accessed on 29 December 2022). The results of Section 3.3 were obtained within the state assignment of Ministry of Science and Higher Education of the Russian Federation (themes "Magnet" No. 122021000034-9 and "Electron" No. 122021000039-4).

Institutional Review Board Statement: Not applicable.

Informed Consent Statement: Not applicable.

Data Availability Statement: The data presented in this study are available on request from the corresponding author.

Acknowledgments: X-ray diffraction and magnetometric measurements were conducted on equipment of the Collaborative Access Center "Testing Center of Nanotechnologies and Advanced Materials" of the Institute of Metal Physics (Ural Branch, Russian Academy of Sciences).

Conflicts of Interest: The authors declare no conflict of interest. The funders had no role in the design of the study; in the collection, analyses, or interpretation of data; in the writing of the manuscript, or in the decision to publish the results.

References

1. Li, L.; Yan, M. Recent progresses in exploring the rare earth based intermetallic compounds for cryogenic magnetic refrigeration. *J. Alloys Compd.* **2020**, *823*, 153810. [CrossRef]
2. Franco, V.; Blázquez, J.S.; Ipus, J.J.; Law, J.Y.; Moreno-Ramírez, L.M.; Conde, A. Magnetocaloric effect: From materials research to refrigeration devices. *Prog. Mater. Sci.* **2018**, *93*, 112–232. [CrossRef]
3. Takeuchi, I.; Sandemana, K. Solid-state cooling with caloric materials. *Phys. Today* **2015**, *68*, 48–54. [CrossRef]
4. Li, L. Review of magnetic properties and magnetocaloric effect in the intermetallic compounds of rare earth with low boiling point metals. *Chin. Phys.* **2016**, *25*, 037502. [CrossRef]
5. Zhang, Y. Review of the structural, magnetic and magnetocaloric properties in ternary rare earth RE_2T_2X type intermetallic compounds. *J. Alloys Compd.* **2019**, *787*, 1173–1186. [CrossRef]
6. Li, Y.; Zeng, Q.; Wei, Q.; Liu, E.; Han, X.; Du, Z.; Li, L.; Xi, X.; Wang, W.; Wang, S.; et al. An efficient scheme to tailor the magnetostructural transitions by staged quenching and cyclical ageing in hexagonal martensitic compounds. *Acta Mater.* **2019**, *174*, 289–299. [CrossRef]
7. Qu, Y.; Cong, D.; Li, S.; Gui, W.; Nie, Z.; Zhang, M.; Ren, Y.; Wang, Y. Simultaneously achieved large reversible elastocaloric and magnetocaloric effects and their coupling in a magnetic shape memory compound. *Acta Mater.* **2018**, *151*, 41–55. [CrossRef]
8. Ouyang, Y.; Zhang, M.; Yan, A.; Wang, W.; Guillou, W.; Liu, J. Plastically deformed La-Fe-Si: Microstructural evolution, magnetocaloric effect and anisotropic thermal conductivity. *Acta Mater.* **2020**, *187*, 1. [CrossRef]
9. Gupta, S.; Suresh, K. Review on magnetic and related properties of RTX compounds. *J. Alloys Compd.* **2015**, *618*, 562–606. [CrossRef]
10. Zhang, H.; Shen, B. Magnetocaloric effects in RTX intermetallic compounds (R = Gd–Tm, T= Fe–Cu and Pd, X = Al and Si). *Chin. Phys.* **2015**, *24*, 127504. [CrossRef]
11. Duy Khanh, N.; Nakajima, T.; Yu, X.; Gao, S.; Shibata, K.; Hirschberger, M.; Yamasaki, Y.; Sagayama, H.; Nakao, H.; Peng, L.; et al. Nanometric square skyrmion lattice in a centrosymmetric tetragonal magnet. *Nat. Nanotechnol.* **2020**, *15*, 444–449. [CrossRef] [PubMed]
12. Welter, R.; Venturini, G.; Malaman, B. Magnetic properties of RFeSi (R≡La–Sm, Gd–Dy) from susceptibility measurements and neutron diffraction studies. *J. Alloys Compd.* **1992**, *189*, 49–58. [CrossRef]
13. Welter, R.; Venturini, G.; Malaman, B.; Ressouche, E. Crystallographic data and magnetic properties of new RTX compounds (R=La–Sm, Gd; T=Ru, Os; X=Si, Ge). Magnetic structure of NdRuSi. *J. Alloys Compd.* **1993**, *202*, 165–172. [CrossRef]
14. Wlodarczyk, P.; Hawelek, L.; Zackiewicz, P.; Roy, T.; Chrobak, A.; Kaminska, M.; Kolano-Burian, A.; Szade, J. Characterization of magnetocaloric effect, magnetic ordering and electronic structure in the $GdFe_{1-x}Co_xSi$ intermetallic compounds. *Mater. Chem. Phys.* **2015**, *162*, 273–278. [CrossRef]
15. Kuchin, A.; Platonov, S.; Gaviko, V.; Yakovleva, M. Magnetic and Structural Properties of $GdFe_{1-x}Ti_xSi$. *IEEE Magn. Lett.* **2019**, *10*, 2509204. [CrossRef]
16. Zhang, H.; Sun, Y.J.; Niu, E.; Yang, L.H.; Shen, J.; Hu, F.X.; Sun, J.R.; Shen, B.G. Large magnetocaloric effects of RFeSi (R=Tb and Dy) compounds for magnetic refrigeration in nitrogen and natural gas liquefaction. *Appl. Phys. Lett.* **2013**, *103*, 202412. [CrossRef]
17. Guzik, A.; Talik, E.; Zajdel, P. Magnetocaloric effect of the $Gd_{3-x}Tb_xCo$ system. *Intermetallics* **2020**, *118*, 106686. [CrossRef]
18. Yang, Y.; Zhang, Y.K.; Xu, X.; Geng, S.H.; Hou, L.; Li, X.; Ren, Z.M.; Wilde, G. Magnetic and magnetocaloric properties of the ternary cadmium based intermetallic compounds of Gd_2Cu_2Cd and Er_2Cu_2Cd. *J. Alloys Compd.* **2017**, *692*, 665–669. [CrossRef]
19. Anisimov, V.I.; Aryasetiawan, F.; Lichtenstein, A.I. First-principles calculations of the electronic structure and spectra of strongly correlated systems. The LDA+U method. *J. Phys. Condens. Matter* **1997**, *9*, 767–808. [CrossRef]
20. Giannozzi, P.; Andreussi, O.; Brumme, T.; Bunau, O.; Buongiorno Nardelli, M.; Calandra, M.; Car, R.; Cavazzoni, C.; Ceresoli, D.; Cococcioni, M.; et al. Advanced capabilities for materials modelling with Quantum ESPRESSO. *J. Phys. Condens. Matter* **2017**, *29*, 465901. [CrossRef]
21. Giannozzi, P.; Baroni, S.; Bonini, N.; Calandra, M.; Car, R.; Cavazzoni, C.; Ceresoli, D.; Chiarotti, G.L.; Cococcioni, M.; Dabo, I.; et al. Quantum ESPRESSO: A modular and open-source software project for Quantum simulations of materials. *J. Phys. Condens. Matter* **2009**, *21*, 395502. [CrossRef]
22. Perdew, J.P.; Burke, J.P.; Ernzerhof, M. Generalized gradient approximation made simple. *Phys. Rev. Lett.* **1996**, *77*, 3865–3868. [CrossRef]
23. Topsakal, M.; Wentzcovitch, R. Accurate projected augmented wave (PAW) datasets for rare-earth elements (RE = La–Lu). *Comput. Mater. Sci.* **2014**, *95*, 263–270. [CrossRef]
24. Quantum ESPRESSO, Pseudopotentials. Available online: https://www.quantum-espresso.org/pseudopotentials (accessed on 29 December 2022).
25. Kuchin, A.G.; Platonov, S.P.; Lukoyanov, A.V.; Volegov, A.S.; Gaviko, V.S.; Mukhachev, R.D.; Yakovleva, M.Y. Remarkable increase of Curie temperature in doped GdFeSi compound. *Intermetallics* **2021**, *133*, 107183. [CrossRef]
26. Napoletano, M.; Canepa, F.; Manfrinetti, P.; Merlo, F. Magnetic properties and the magnetocaloric effect in the intermetallic compound GdFeSi. *J. Mater. Chem.* **2000**, *10*, 1663–1665. [CrossRef]
27. Ślaski, M.; Szytuła, A.; Leciejewicz, J.; Zygmunt, A. Magnetic properties of $RERu_2Si_2$ (RE = Pr, Nd, Gd, Tb, Dy, Er) intermetallics. *J. Magn. Magn. Mater.* **1984**, *46*, 114–122. [CrossRef]

28. Gębara, P.; Hasiak, M. Determination of Phase Transition and Critical Behavior of the As-Cast GdGeSi-(X) Type Alloys (Where X = Ni, Nd and Pr). *Materials* **2021**, *14*, 185. [CrossRef] [PubMed]
29. Mukhachev, R.D.; Lukoyanov, A.V. Composition-Induced Magnetic Transition in $GdMn_{1-x}Ti_xSi$ Intermetallic Compounds for x = 0–1. *Metals* **2021**, *11*, 1296. [CrossRef]
30. Baglasov, E.D.; Lukoyanov, A.V. Electronic Structure of Intermetallic Antiferromagnet GdNiGe. *Symmetry* **2019**, *11*, 737. [CrossRef]

MDPI AG
Grosspeteranlage 5
4052 Basel
Switzerland
Tel.: +41 61 683 77 34

Metals Editorial Office
E-mail: metals@mdpi.com
www.mdpi.com/journal/metals

www.ingramcontent.com/pod-product-compliance
Lightning Source LLC
Chambersburg PA
CBHW041953260926
51850CB00019B/2696